Water Resou

T0252314

SOIL AND GROUNDWATER CONTAMINATION: NONAQUEOUS PHASE LIQUIDS – PRINCIPLES AND OBSERVATIONS

Alex Mayer
S. Majid Hassanizadeh
Editors

R.W. Falta, S. M. Hassanizadeh,
T.H. Illangasekare, I. Javandel,
K. H. Jensen, A. S. Mayer, and M. Oostrom
Contributing Authors

American Geophysical Union
Washington, DC

Mayer, Alex.
  Soil and groundwater contamination : nonaqueous phase liquids / Alex Mayer.
    p. cm. -- (Water resources monograph ; 17)
  Includes bibliographical references and index.
    1. Oil pollution of soils. 2. Oil pollution of water. 3. Nonaqueous phase liquids--
Environmental aspects. 4. Soil remediation. 5. Groundwater--Purification. I. Title.
II. Series.

  TD879.P4M353 2005
  628.1'6833--dc22

                          2005006928

ISBN-13 : 978-0-87590-321-7
ISBN-10 : 0-87590-420-3
ISSN 0170-9600

Table of Contents

Preface

Awareness of the problem of soil and groundwater contamination by nonaqueous phase liquids (NAPLs) began in the late 1970s to early 1980s. This awareness grew out of observations that sites, whose remediation was predicted to conclude over several years, were nowhere near meeting cleanup goals, even after decades of remediation efforts. Many such sites were found to be contaminated with essentially immiscible organic liquids, such as petroleum hydrocarbons and chlorinated solvents. It was eventually recognized that, due to their low solubility, small amounts of these liquids had the potential to contaminate very large volumes of soils and groundwater. Furthermore, it was recognized that the uneven distribution of the liquids in the subsurface, coupled with the low solubility and relatively low flow rates of groundwater, were giving rise to excessive time frames for remediation with typical cleanup goals.

During the decades that followed these "discoveries," intensive research efforts were dedicated toward investigating the flow, transport, and interphase mass exchange of NAPLs. These efforts have led to an enhanced understanding of NAPL migration in the vadose and saturated zones, improved characterization of NAPLs as sources of groundwater contamination, appropriate site investigation techniques for assessing NAPL contamination, and better technologies and strategies for remediating NAPL-contaminated sites. In particular, porous media heterogeneity has been recognized as perhaps the single most important factor in determining NAPL distributions, the subsequent function of NAPLs as a source of groundwater contamination, and the limiting factor in remediating NAPL-contaminated sites.

In this book, we present overviews of recent advancements within the context of a more complete reference on the principles of NAPL migration and distribution for engineering and scientific consultants, academics, and students. We begin by explaining the fundamental physical and chemical phenomena that impact NAPL transport in the subsurface. The book then takes a tutorial approach, where we pose a probable field scenario, and then explain the phenomena that may have created the scenario with appropriate discussion on the fundamental physical and chemical phenomena of concern. The book introduces the most commonly used

Soil and Groundwater Contamination: Nonaqueous Phase Liquids
Water Resources Monograph 17
Copyright 2005 by the American Geophysical Union
10.1029/17WM00

equations for describing NAPL fate and transport in the subsurface, followed by numerical examples of the application of these equations, and offers multiple illustrations (with well over one hundred figures) to further explain the concepts described in the text. Numerous examples of the application of mathematical formulas and analyses to assess NAPL contamination are given as well. The book is supplemented by a user-friendly CD that contains spreadsheets used in many of the example calculations, color versions of some of the illustrations, and movies illustrating NAPL migration.

This book is the result of collaborative effort between the editors and several contributors with a wide range of experience in teaching and in solving groundwater contamination problems in the laboratory and the field. The editors outlined the book, solicited sections from the contributors, and edited the contributions to produce an integrated book.

Co-editor Dr. Alex Mayer is a Professor in the Department of Geological & Mining Engineering & Sciences, Michigan Technological University. Dr. Mayer's Ph.D. in Environmental Engineering is from the University of North Carolina at Chapel Hill, 1992. In addition to his role as co-editor, Dr. Mayer contributed to Chapter 1 and Sections 3.5, 4.2, 5.1, and 5.2. Co-editor Dr. S. Majid Hassanizadeh is Professor of Hydrogeology in the Department of Earth Sciences of Utrecht University, The Netherlands. He received his Ph.D. from Princeton University in 1979. Also in addition to co-editing, Dr. Hassanizadeh contributed to Chapter 1 and Section 5.1.

The contributors and their contributed sections include the following. Dr. Ron Falta is a Professor of Geology and Environmental Engineering at Clemson University, South Carolina. Dr. Falta received his Ph.D. in Mineral Engineering from the University of California, Berkeley in 1990. Dr. Falta contributed to Sections 2.3 and 5.4. Dr. Tissa Illangasekare is AMAX Distinguished Chair of Environmental Science and Engineering and Professor of Civil Engineering at the Colorado School of Mines. He received his Ph.D. in Civil Engineering from Colorado State University in 1978. Dr. Illanagasekare contributed to Sections 3.2, 3.3, 3.4, and 3.5. Dr. Iraj Javandel is a senior scientist at Lawrence Berkeley National Laboratory at the University of California. He received his Ph.D. in Civil Engineering/Hydrogeology from the University of California at Berkeley in 1968. Dr. Javandel contributed to Sections 3.5 and 5.3. Dr. Karsten H. Jensen, M.Sc. and Ph.D. in Hydrology from the Technical University of Denmark, is currently professor at the Geological Institute at the University of Copenhagen. Dr. Jensen contributed to Sections 2.1, 2.2, 3.1, 3.2, and 3.3. Dr. Mart Oostrom has been employed at Pacific Northwest National Laboratory for almost 10 years, where he directs the Multifluid Flow Research Laboratory. Dr. Oostrom received his Ph.D. from Auburn University, Alabama, in 1991. Dr. Oostrom contributed to Sections 4.1 and 5.2.

Dr. Mayer wishes to acknowledge the support of the Fulbright U.S. Scholar Program and a Visiting Professorship at the Delft University of Technology, which were instrumental in the creation of this book. This book is based on lectures given at the advanced course "Contamination of soil and groundwater by nonaqueous phase liquids (NAPLS)-Problems and solutions," 1-3 July 2002, Delft, The Netherlands. The course was organized by Postgraduate school PAO of Delft University of Technology. The editors are grateful to David Oostveen, M.Sc. student at Faculty of Civil Engineering and Geosciences in Delft for his assistance with word processing aspects such as referencing and indexing. We are also grateful to the anonymous reviewer, whose suggestions were extremely helpful. Finally, we wish to thank the staff at AGU books for their considerable expertise in publishing and for seeing this project through to completion.

The editors of this book wish to dedicate this book to our beloved companions: Suzanne and Forooz.

Alex Mayer
Department of Geological & Mining Engineering & Sciences
Michigan Technological University

S. Majid Hassanizadeh
Department of Earth Sciences
Utrecht University, The Netherlands

Introduction

A significant portion of contaminated soil and groundwater sites contains non-aqueous phase liquids (NAPLs). NAPLs are hazardous organic liquids that are immiscible with water and form a visible, separate oily phase in the subsurface. Their migration is governed by gravity, viscous forces, and capillary forces. If NAPLs were truly insoluble in water, their impact on groundwater quality would be very limited. But, NAPL components can dissolve in water in very small amounts (yet much higher than drinking water limits) and at very low rates. As a result, given the high toxicity of NAPL components, a small volume of NAPL in soil can form a long-term threat to the groundwater quality. Thus, unless properly managed, NAPLs can exist in the subsurface for decades and can contaminate large volumes of groundwater. However, it is very difficult to design effective remediation schemes, due to the complex behavior of NAPLs in the subsurface. In fact, the presence of NAPL has been shown to be a significant limiting factor in site remediation [EPA, 2003]. This is partly due to hydrogeologic factors, such as complex heterogeneity patterns, and the presence of low permeability zones. But more importantly, and tied with hydrogeologic factors, it is because of the very complex nature of the various processes that affect the migration of NAPLs and transport of their dissolved components.

During the past few decades, a huge body of literature has developed on the occurrence of NAPL in the subsurface, on the spread of NAPL as a separate phase and/or transport of its components by gas and water phases, and on ways of containing and cleaning the contamination. We do not intend to give a review of the literature in this monograph. But, we will make extensive references to the published literature throughout the book. Here we give a very short list of various categories of literature that have been consulted in this monograph. First, there is a myriad of papers that have appeared in journals and conference proceedings. Also, national laboratories, agencies, and institutes have contributed extensively to the body of literature in this area. In particular, publications by the American Petroleum Institute, API, have treated various aspects of NAPL pollution and remediation (see, e.g., API [1986]; Charbeneau et al. [1999] Huntley and Beckett [2002b], [2004]; Sale [2001]; or the API LNAPL web site, http://groundwater.api.org/lnapl/). Also, the U.S. Environmental Protection Agency (EPA) has published a large number of reports on NAPL in the subsurface (see, e.g., EPA [2003]; Parker et al. [1995]; Sabatini et al. [1996];

Soil and Groundwater Contamination: Nonaqueous Phase Liquids
Water Resources Monograph 17
Copyright 2005 by the American Geophysical Union
10.1029/17WM01

Wilson et al. [1990]; or the EPA CLU-IN web site http://clu-in.org/). EPA has also published outreach documents, which explain pollution and remediation issues in a simple language (see, e.g., "The Citizen's Guide" series, http://clu-in.org/products/ citguide/). A number of specialized textbooks include sections on NAPL flow and transport in soil and groundwater (see, e.g., Bedient et al. [1999], Corey [1994], Fetter [1999], Helmig [1997], Pankow and Cherry [1996]). There are a few books that deal specifically with NAPL remediation methods. Examples are Nyer et al. [1996] and Simpkin et al. [1999]. A publication of particular interest for remediation issues is the monograph "Alternatives for Groundwater Cleanup" by the National Research Council [1994].

The purpose of this book is to provide the background for solving practical problems concerning NAPL contamination and remediation. The book discusses principles of multiphase flow, transfer of NAPL components to water and gas phases, and their transport in the subsurface. Where a principle is introduced, its significance to practical problems, such as site investigations and remediation, is discussed. Special attention is paid to the role of heterogeneities in the spatial distribution of NAPLs. Concepts are illustrated with the aid of graphs, drawings, and photographs. The book describes well-established concepts, but also the gaps in our understanding of NAPLs at the field scale, illustrated with examples from field, laboratory, or modeling studies. Many situations that at first seem to be counterintuitive are described, and then are explained with the aid of basic principles. Examples of counterintuitive situations include the presence of LNAPL (lighter-than-water NAPL) below groundwater table or upgradient occurrence of dissolved NAPL components.

This book will be useful to a wide range of audiences. It can provide practicing engineers and scientists who are involved in NAPL contamination and remediation studies with the necessary background. It can serve as a reference on multiphase flow and transport phenomena for professionals (e.g. consultants and regulators) who are concerned with evaluating or remediating NAPL-contaminated sites. It can be used as a supplement to groundwater or hydrogeology course work for upper-level undergraduate or beginning graduate students. It is assumed that the reader has a background equivalent to an undergraduate course in quantitative groundwater hydrology.

This book consists of four main chapters. In Chapter 2, basic principles of multiphase flow and contaminant transport in porous media are presented in a concise manner. Properties and parameters responsible for the migration and/or trapping of NAPL are described. Various interphase mass transfer processes are introduced and multicomponent transport mechanisms are explained. In Chapter 3, the interplay of hydrogeologic factors with NAPL characteristics is described in order to explain complex patterns of NAPL distribution in the subsurface. Phenomena such as pooling, trapping under the water table, permeability barriers, unstable fronts, and fingering are described and explained. In Chapter 4, common observations associated with monitoring and assessment of sites contaminated with NAPLs are described, including observations of NAPLs in monitoring wells and soil samples and observations of NAPLs dissolved in groundwater. Finally, Chapter 5 deals with the remediation of NAPL-polluted sites. Attention is focused on issues and phenomena associated with three standard and widely-applied methodologies. These are: hydraulic removal of NAPL (separate phase removal of LNAPL via wells or trenches), pump-and-treat (extraction of groundwater containing dissolved NAPL components via wells) and soil vapor extraction (extraction of gas phase containing volatilized NAPL components).

This book is based on lectures given at the advanced course "Contamination of soil and groundwater by nonaqueous phase liquids (NAPLS)-Problems and solutions," 1-3 July 2002, Delft, The Netherlands. The course was organized by Postgraduate school PAO of Delft University of Technology. The editors are grateful to David Oostveen, M.Sc. student at Faculty of Civil Engineering and Geosciences in Delft for his assistance with word processing aspects such as referencing and indexing. We are also grateful to the anonymous reviewer, whose suggestions were extremely helpful.

Fundamentals

2.1   NAPL CHARACTERISTICS AND BEHAVIOR OF NAPLS IN THE SUBSURFACE

Summary: Nonaqueous phase liquids (NAPLs) move as a separate phase in the subsurface. An evaluation of the migration and distribution of the NAPL phase is generally the first step in a contamination impact assessment. The density of the NAPL has a major impact on the migration pattern. A light NAPL tends to float on the water table while a dense NAPL may penetrate deep into an aquifer, leading to very different migration scenarios. A trace of residual NAPL is left behind along the migration pathways. This fraction may partition to the water or air phases and thus serve as a long-term source of contamination.

Organic fluids with low solubilities in water are generally referred to as nonaqueous phase liquids (NAPLs). NAPLs move as a separate phase and are often visible to the naked eye, as opposed to soluble contaminants that travel with the groundwater. NAPLs have been widely used in various industries and are known to be present at numerous industrial and waste disposal sites and at many unknown locations where they have been spilled either accidentally or on purpose. Due to their potential toxicity and widespread occurrence, they constitute a serious environmental problem. It is important to have a thorough understanding of the many processes that affect their behavior and fate in the subsurface in order to assess their potential impacts and to design effective control measures.

NAPLs have different chemical and physical characteristics; thus their behavior and fate in the subsurface can be very different. The density of these contaminants, in particular, has a decisive influence on the migration in the subsurface. It is common to group NAPLs in two classes, based on their density:

LNAPL: Light liquids with a density less than water, implying that they tend to float and move along the top of the saturated groundwater zone. (Examples: gasoline, benzene, xylene)

Soil and Groundwater Contamination: Nonaqueous Phase Liquids
Water Resources Monograph 17
Copyright 2005 by the American Geophysical Union
10.1029/17WM02

DNAPL: Dense liquids with a density higher than water, implying that they have
         a tendency to penetrate the water table and move deeper into the aquifer
         system. (Examples: chlorinated solvents like TCE, TCA and PCE)

A NAPL may be a pure chemical or a mixture of constituents with different
chemical characteristics. When present in the subsurface, partitioning of NAPL
components takes place and some of the constituents may dissolve in the water
phase or exist as vapor in the air phase. Thus, besides moving as a separate fluid
phase, NAPLs may also be transported as soluble components with the flowing
groundwater and as volatile components in the air phase. Although concentrations
of organic compounds in water and air phases are generally very small; due to
their high toxicity, they may exceed the allowable concentrations in groundwater
or indoor air. From an environmental hazard point of view, the migration of
organic compounds in both water and air is crucial [Pankow and Cherry, 1996].

In this section, we address the first step in NAPL contamination scenarios: the
migration and distribution of the NAPL. Figure 2.1a provides a schematic
illustration of the migration of an LNAPL in the unsaturated zone following a
spill above a water table. In the unsaturated zone, the contamination moves predominantly in the vertical direction under the influence of gravity. Lateral spreading also occurs due to the effect of capillary forces. During migration in the
unsaturated zone, the LNAPL predominantly displaces air, since the air phase has
a high mobility due to its low density and viscosity. Some displacement of capillary water also may take place, producing a water front ahead of the NAPL front.
In the unsaturated zone, some of the contamination is left behind as residual
NAPL due to a combination of capillary trapping and immobilization caused by
disconnection. The residual NAPL serves as a long-term source of contamination
through infiltrating precipitation or a fluctuating water table.

If the LNAPL spill is sufficiently large, it may eventually reach the capillary
fringe zone and the water table. The LNAPL first starts accumulating on the capillary fringe zone. Initially, NAPL is under negative pressure (relative to atmospheric
pressure). However, as more NAPL builds up, the pressure becomes positive and a
"head" of NAPL develops on the capillary fringe. If the volume of the spill is large
enough, the NAPL head typically becomes high enough to overcome the buoyancy
and capillary forces, displace water from the capillary and saturated zone, and
penetrate the water table beneath the spill location. Concurrently, lateral spreading
takes place predominantly in the direction of the flowing groundwater. Away from
the spill location, in the downstream groundwater flow direction, the NAPL lens
gradually thins. Once the NAPL spill volume is exhausted and the downward, vertical flow of the NAPL begins to cease, the NAPL that has penetrated the saturated
zone is subsequently displaced by water. However, some NAPL will be left behind
in the saturated zone due to capillary trapping effects.

The spreading behavior illustrated in Figure 2.1a is only applicable for idealized
geological conditions. The configuration of the NAPL will depend on a variety of
factors, most notably the spill history, liquid properties and the geological setting.
We see in the inset picture in Figure 2.1a that the LNAPL may be found as a

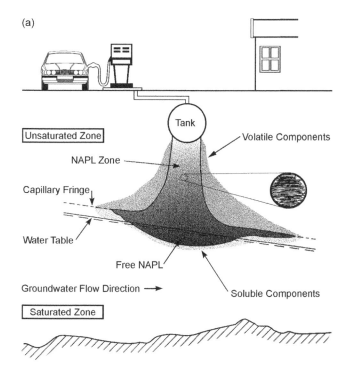

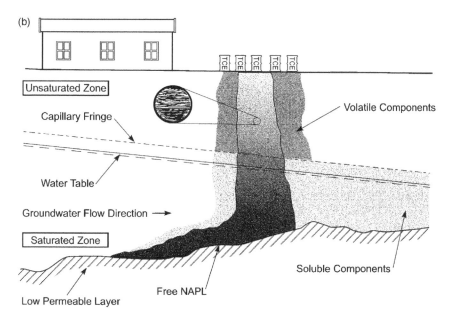

Figure 2.1b. Idealized migration of (a) an LNPL plume and (b) a DNAPL plume (after Abriola [1984]).

series of thin, interconnecting lenses or pools that follow the subtle laminations of the porous media.

The migration of a DNAPL spill, in the unsaturated zone, where the NAPL displaces predominantly air and typically only minor amounts of water, as shown in Figure 2.1b, follows a spreading behavior similar to an LNAPL spill. Due to different density and viscosity characteristics, a DNAPL may be less prone to horizontal spreading and take a more direct pathway towards the saturated zone. The most important difference between the two types of liquids occurs when the DNAPL reaches the saturated zone. If the spill is of such quantity that the displacement pressure (see Section 2.2.2) is exceeded, the DNAPL penetrates the groundwater zone and migrates further downwards. The most important controlling factor for the further migration of a DNAPL is the presence of less permeable boundaries that arrest or divert the vertical migration. Also for DNAPLs, some fluid is retained, as residual saturation, in both the unsaturated and saturated zones by capillary forces. As in the case of the schematic picture in Figure 2.1b, we see in the inset that the DNAPL may be found as a series of pools that follow the laminations of the porous media. The pools may be interconnected by of fingers of DNAPL that have managed to penetrate lower permeability materials.

For both LNAPLs and DNAPLs, it is worth noting that their behaviors are significantly different in the unsaturated and saturated zones. The difference in behavior is caused by differences in "wettability," a concept that is discussed in Section 2.2.1. In the unsaturated zone, the NAPL is typically an intermediate wetting fluid, whereas it is typically the non-wetting fluid in the saturated zone.

Table 2.1 lists physicochemical parameters of importance for immiscible fluid flow for some selected compounds. For reference, the properties of water and air are listed as well. The relative mobility to water is a convenient parameter for determining the ability of a liquid to migrate in the subsurface and is defined as

TABLE 2.1.   Physicochemical parameters for water, air and selected LNAPLs (light gray shading) and DNAPLs (dark gray shading) compounds for the temperature range 20-25°C [Davis, 1997; Mercer and Cohen, 1990].

	Density (g/cm^3)	Viscosity (cP)	Relative Mobility to Water	Boiling Point (°C)	Solubility in Water (mg/l)	Vapor pressure (mm Hg)
Water	1.00	1.00	1.00	100	–	24
Air	0.0011	0.0018	0.61	–	–	–
Toluene	0.86	0.55	1.39	111	535	28
Benzene	0.88	0.60	1.32	80	1750	95
o-Xylene	0.88	0.81	1.11	144	175	7
TCE	1.46	0.55	2.39	87	1100	58
TCA	1.35	0.90	1.50	74	1500	123
PCE	1.61	0.84	1.73	121	~200	17

$$\Gamma = \frac{\rho_{nw} / \mu_{nw}}{\rho_w / \mu_w} \tag{2.1}$$

where ρ_{nw} and ρ_w are the densities of NAPL and water, respectively, and μ_{nw} and μ_w are the dynamic viscosities of NAPL and water, respectively. The mobility of DNAPLs shown in Table 2.1 is about twice that of water, making these liquids extremely mobile in the subsurface.

2.2 PARAMETERS RESPONSIBLE FOR NAPL DISTRIBUTION AND FLOW IN THE SUBSURFACE

Summary: In a porous medium where two or more immiscible phases are present at the same time, cohesive forces (attractive forces between alike molecules) and adhesive forces (attractive forces between different molecules) in combination give rise to capillary forces that are responsible for the retention of fluids. Capillary forces are manifested by a pressure difference across the interface between two immiscible phases, referred to as capillary pressure, which is a function of the content of the phases. The relationship between capillary pressure and phase content is referred to as the capillary pressure curve. This curve, and the relative permeability curve describing the relationship between relative permeability and phase content, represent the two fundamental functional relationships, which need to be specified when solving multiphase flow problems. The mathematical framework for describing the simultaneous flow of immiscible phases in a multiphase system is developed from an extended form of Darcy's law for groundwater flow.

2.2.1 *Interfacial Tension and Wettability*

Consider a porous medium whose pore spaces are occupied by two immiscible phases, e.g. water and air. At the pore scale, a sharp interface is present between the two phases and one of the phases typically wets the grain surfaces more readily than the other. This situation reflects two important phenomena: interfacial tension and wettability. Within each fluid phase, there are attractive forces between each molecule. These forces, referred to as *cohesive* forces, are manifested by a material tending to hold itself together. In addition, there are intermolecular forces between molecules in the separate fluid phases, referred to as *adhesive* forces, which are manifested as the tendency of the fluids to cling to each other. If we examine the simplified picture shown in Figure 2.2, a water molecule sufficiently far enough away from the air-water interface is, on average, uniformly attracted to surrounding molecules. There is a zero net balance of cohesive forces on this molecule.

In contrast, a molecule residing at the interface is subject to both cohesive forces from below and adhesive forces from above. In the case of air and water, the

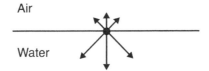

Air

Water

Figure 2.2. Imbalance of cohesive forces at the interface between two immiscible phases (after Charbeneau [2000]).

cohesive forces within the fluids are much stronger than the adhesive forces between the fluids. As a consequence, there is a net imbalance of forces acting on the molecules at the interface. The force imbalance suggests that, in order to bring a molecule towards the interface, work must be performed to overcome the net downward force. This work is equivalent to imparting a potential energy to the molecule.

The molecules at the interface thus possess an additional energy compared to the molecules in the bulk water and this excess surface energy between the two phases is referred to as interfacial tension (force per unit length). This force acts tangentially to the interface separating the two phases and behaves similar to a stretched membrane. Values of interfacial tensions for a few selected compounds are listed in Table 2.2. Note that in this analysis, collision forces that result in a fluid's internal pressure have not been taken into account.

When two phases are brought into contact with the porous medium, they compete against each other for the surfaces of the grain particles. One of the phases

TABLE 2.2. Typical values for interfacial tensions for the temperature range 20-25°C [Davis, 1997; Mercer and Cohen, 1990].

Compound	Fluid–Air Interfacial Tension (dyne/cm)	Fluid–Water Interfacial Tension (dyne/cm)
Water	72.0	–
Toluene	27.9	36.1
Benzene	28.9	35.0
o-Xylene	36.1	30.3
TCE	29.5	34.5
TCA	30.8	45.0
PCE	32.9	44.4

exhibits a stronger affinity for the grain particles than the other and thus coats the surfaces of the grain particles. The other phase is expelled towards the interior of the pore spaces. Differences in adhesive forces (attractive forces between different types of molecules) are responsible for the greater or lesser affinity of one phase for the grain surfaces. We use the term wettability to describe the tendency of one phase being attracted to the grain particles in preference to another phase [Imhoff and Miller, 2000]. For two-phase systems we refer to the phases as follows:

Wetting phase: the phase that shows the greatest preference towards contact with the grain particles

Non-wetting phase: the phase that shows the least preference towards contact with the grain particles

The wettability of a solid surface towards two phases can be measured by observing the contact angle of a droplet of a test fluid when placed on the surface and surrounded by a background reference phase. The contact angle,θ, is defined as the tangent to the droplet measured through the test fluid at the intersection between the two immiscible phases and the solid surface (see Figure 2.3). We note from Figure 2.3 that the contact angle of the fluid of interest with respect to a solid surface depends on the fluid surrounding the fluid of interest. A contact angle of $\theta < 90°$ indicates that the test fluid is wetting phase (with respect to the surrounding fluid); an angle of zero indicates that the test fluid is perfectly wetting; an angle $\theta > 90°$ indicates that the test fluid is the non-wetting phase (with respect to the surrounding fluid); and an angle of 90° (within some tolerance levels) suggests a neutral-wetting system. In general, if the adhesive forces between the solid and the test fluid is greater than the cohesive forces in the fluid, the contact angle is $\theta < 90°$ and the test fluid acts as the wetting fluid [Bedient et al., 1999].

In practical field applications, it may not be feasible to apply these principles to determine wettability relationships between water, NAPL, and air. However, wettability orders listed in Table 2.3 are applicable as a first approximation. As shown in Table 2.3, it is most common that grains in natural subsurface systems are water wet. Exceptions to this rule may occur in (1) very dry unsaturated systems, in which case a NAPL release may lead to a NAPL wet system; (2) carbonate materials, which are often preferentially wetted by hydrocarbons; and (3) soils of a high organic content, which may also form a NAPL wet system. In addition,

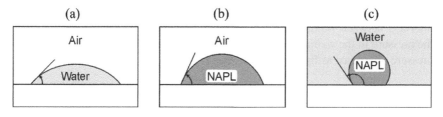

Figure 2.3. Typical wettability preferences for combinations of water, NAPL and air (after Bedient et al., [1999]).

TABLE 2.3.   Typical wettability orders for different phase combinations [Bedient et al., 1999].

Two-phase system	Wetting phase	Non-wetting phase
Water – Air	Water	Air
NAPL – Air	NAPL	Air
Water – NAPL	Water	NAPL

Three-phase system	Wetting phase	Intermediate wetting phase	Non-wetting phase
Water – NAPL – Air	Water	NAPL	Air

systems of mixed wettability, where individual grains or even portions of grain surfaces vary in wettability, may occur.

The wettability order has a very strong influence on the distribution of phases and thus also on the mobility and retention characteristics of the NAPL phase. This influence is illustrated in Figure 2.4, which shows NAPL distributions expected for water-wet and NAPL-wet systems. For water-wet systems, the water phase will envelope the grains and the NAPL phase will be present in the largest pores. In contrary, for a NAPL-wet system, the NAPL phase will reside next to the grains and occupy the smallest pores.

When three phases are present at the same time, it is usually assumed that water is the wetting fluid and NAPL is the intermediate wetting fluid. According to this wettability order, we may visualize the phase distributions at the pore scale as shown schematically in Figure 2.5. Water envelopes all the grain particles and forms a continuous phase; NAPL forms a film around the air phase (eliminating the contact between water and air); and air is present in the interior of the pore spaces.

2.2.2   Capillary Pressure and Capillary Pressure Curves

Due to the combined effect of cohesive and adhesive forces, a curved interface will appear between two immiscible phases coexisting in pore spaces and a pressure difference exists across the interface. As indicated by the curvature of the interface between fluids, the pressure in the non-wetting phase is larger than the pressure in the wetting phase. Following standard conventions, we define the pressure difference as the capillary pressure, p_c:

$$p_c = p_{nw} - p_w \qquad (2.2)$$

where p_{nw} is the non-wetting phase pressure and p_w is the wetting phase pressure. According to this definition, the capillary pressure is a positive quantity.

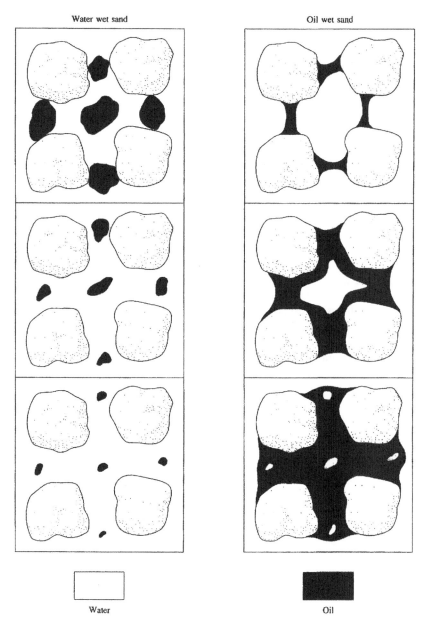

Figure 2.4. Different fluid distributions for water-wet and NAPL-wet porous media containing water and NAPL (from Fetter [1999]). Fetter, C.W., Contaminent Hydrogeology, 2nd Edition, 1999. Reprinted by permission of Pearson Education, Inc., Upper Saddle River, NJ.

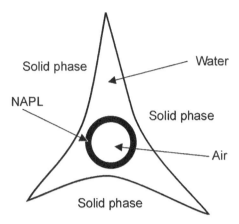

Figure 2.5. Illustration of the idealized distribution of water, NAPL, and air at the pore scale in a porous medium (from Hofstee et al. [1997]). Reprinted from Journal of Contaminant Hydrology, Vol. 25, C. Hofstee, J.H. Dane and W.E. Hill, Three-fluid retention in porous media involving water, PCE and air, 235-247 Copyright 1997, with permission from Elsevier.

Figure 2.6 shows a cross section through a pore where an interface between two immiscible phases exists. At equilibrium conditions, the pressure difference across the interface is balanced by the interfacial tension of the interface as given by Laplace's equation of capillarity:

$$p_c = \frac{2\sigma}{r'} \tag{2.3}$$

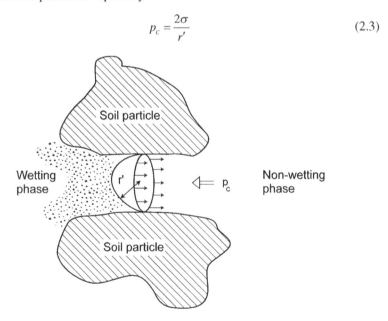

Figure 2.6. Idealized interface between two immiscible phases in the void between two grain particles.

where σ is the interfacial tension, and r' is the radius of curvature for a hemispherical interface. This equation states that capillary pressure is proportional to the interfacial tension and inversely proportional to the radius of curvature of the interface. One result of this relationship is that a threshold capillary pressure (referred to as a displacement or bubbling pressure) is required to drive a non-wetting phase through pores. According to Equation (2.3), this threshold pressure is larger for smaller pore sizes.

Example: Calculate the capillary pressure p_c across a water-air interface where the pressure in the water phase p_w is 0.8×10^5 Pa and pressure in the air phase p_a is at atmospheric pressure.

Solution: Using Equation (2.2) and assuming the atmospheric pressure to be $1.01 \cdot 10^5$ Pa the capillary pressure is calculated to be

$$p_c = p_a - p_w = 1.01 \cdot 10^5 \text{ Pa} - 0.8 \cdot 10^5 \text{ Pa} = 0.21 \cdot 10^5 \text{ Pa}$$

Example: Calculate the capillary pressure across a water-air interface with a radius of curvature, r', of 10 μm.

Solution: Using Equation (2.3) and employing a value for the interfacial tension of 72.0 dyne/cm (Table 2.2) the capillary pressure is calculated to be

$$p_c = \frac{2\sigma}{r'} = \frac{2 \times 72 \times 10^{-3} \text{N/m}}{10 \times 10^{-6} \text{m}} = 0.14 \times 10^5 \text{Pa}$$

For practical scales of porous media systems, it is obviously infeasible to deal with fluid configurations and fluid flow at the pore scale. Instead, the continuum assumption is invoked, implying that the grain skeleton and each fluid phase can be considered as separate, overlapping continua. In order to define a macroscopic capillary pressure (and other relevant, macroscopic hydraulic properties), spatial averaging is made over a number of grain particles or pore volumes (see, e.g., Corey [1994]).

At the continuum scale, we can argue that there exists a relationship between capillary pressure and the volumetric contents of the phases in the porous medium. First, we must recognize that the porous medium consists of a distribution of pores with different radii. Now, consider applying a macroscopic capillary pressure to a porous media accessible to both wetting and non-wetting fluid. As we increase the capillary pressure, the non-wetting fluid would invade the larger pores and the wetting fluid would be present in smaller pores. The larger sized pores could not support the capillary pressure and would release the wetting fluid. Thus, generally speaking, the larger the capillary pressure, the smaller the wetting phase amount will be.

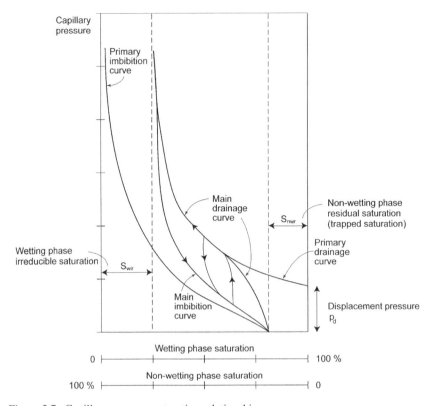

Figure 2.7. Capillary pressure-saturation relationship.

The relationship between capillary pressure and the wetting phase content is referred to as the capillary pressure curve, which is one of the basic functional relationships characterizing multiphase flow. Figure 2.7 shows a typical capillary pressure-wetting fluid saturation relationship for a given porous medium. The ordinate axis represents capillary pressure and the abscissa axis shows the wetting phase saturation, S_w, defined as

$$S_w = \frac{\theta_w}{\phi} \tag{2.4}$$

where θ_w is the wetting phase content and ϕ is the porosity. In contaminant hydrogeology, it is sometimes convenient to express capillary pressure in terms of an equivalent height of a water column, or capillary pressure head, defined as

$$h_c = \frac{p_c}{\rho_w g} \tag{2.5}$$

Example: Calculate the capillary pressure head for a capillary pressure of 0.21·10⁵ Pa.

Solution: Using Equation (2.5) the capillary pressure head is calculated to be

$$h_c = \frac{p_c}{\rho_w g} = \frac{0.21 \cdot 10^5 \, \text{Pa}}{1000 \text{kg} / \text{m}^3 \cdot 9.81 \text{m} / \text{s}^2} = 2.14 \text{m}$$

Sale [2001] has summarized the laboratory techniques for determining capillary pressure curves. The laboratory determination of capillary pressure curve provides a good setting for further explaining the relationship between pressure and saturation. A typical experiment (see Figure 2.8), begins with a porous medium sample that is fully saturated by the wetting fluid. The sample is gradually drained by increasing the capillary pressure at incremental steps, resulting in displacement of the wetting fluid by the non-wetting phase. At each increment, the volume of wetting fluid remaining in the sample is determined. The resulting curve, representing corresponding values of wetting phase saturation and capillary pressure at equilibrium conditions, is known as the primary drainage curve. Figure 2.7 shows that, eventually, a further increase in capillary pressure will not lead to any further displacement of the wetting phase fluid due to phase fragmentation or strong wetting phase attachment to the grain particles. The saturation at which this condition occurs is known as the irreducible wetting phase saturation, S_{wir}.

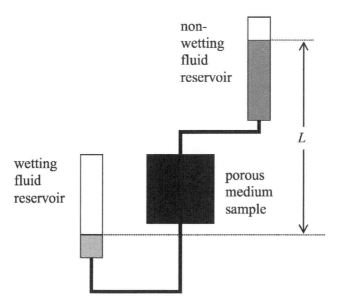

Figure 2.8. Simple experimental apparatus for determining the capillary pressure curve. The capillary pressure is increased by increasing the distance L between the two fluid levels.

After drainage to irreducible wetting phase saturation, the sample can be rewetted again. By plotting the experimental results for this experiment, the main imbibition curve is obtained (see Figure 2.7). As shown, full wetting phase saturation will not occur at zero capillary pressure. Some of the non-wetting phase will be trapped as isolated bubbles in the largest pores and thus will not be displaced. This saturation value is referred to as trapped or residual non-wetting phase saturation, S_{nwr}, and is discussed in Section 3.1.

The primary drainage and main imbibition curves do not coincide because the pores in the porous medium wet and drain differently. The non-unique relationship between capillary pressure and saturation is referred to as hysteresis. If another drainage experiment is made after imbibition, the main drainage curve is obtained. The main drainage and imbibition curves envelope the capillary pressure-saturation relations that occur in practice. The hysteresis phenomenon is not restricted to the main drainage and imbibition curves. An infinite number of scanning curves may occur depending on at what point the drainage or imbibition process is reversed.

Figure 2.7 also shows the primary imbibition curve that applies when imbibing into a medium completely saturated by non-wetting fluid. Like the primary drainage curve, this curve is not relevant for practical applications.

Another important characteristic of the capillary pressure curve shown in Figure 2.7 is that drainage of an initially fully saturated porous medium does not occur until some threshold value of capillary pressure is exceeded. This value is called the displacement pressure, which also is an important characteristic from a contamination viewpoint because the capillary pressure must exceed the displacement pressure before a NAPL can enter a water-saturated layer.

Figure 2.9 shows some examples of capillary curves for two different porous materials and for two different two-phase mixtures. As shown in Figure 2.9, the fine-grained sand composed of smaller pores has a higher capillary retention capability (higher wetting phase saturation for given capillary pressure) than the medium-grained sand with larger pores. The capillary pressure curves also depend on the phase combination. Figure 2.9 shows that a water-air system has a higher retention than a TCE-air system because the interfacial tension for the water-air combination is higher than for the TCE-air combination.

Since the capillary pressure-saturation relationship is non-unique due to hysteresis, a proper description requires specification of a family of curves. In practical applications, however, it is common to simplify the description and assume that the capillary pressure curve can be characterized by a single curve. This simplification is introduced because it is experimentally difficult and tedious to establish the complete family of curves. It is typical to represent the capillary pressure relationship by the primary or the main drainage curve, simply because most experiments establish this particular curve.

It is often convenient, particularly in numerical modeling applications, to fit a parametric function to the capillary pressure curve measurements. Several parametric models have been suggested in the literature, but the two most widely used

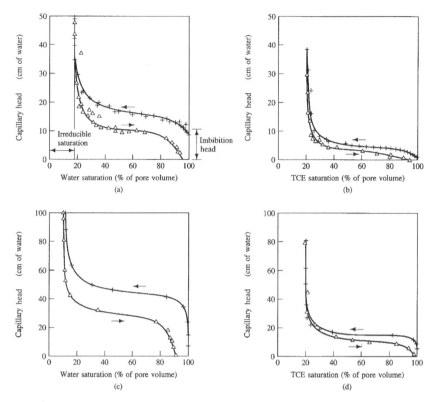

Figure 2.9. Capillary head curves for (a) water-air in medium-grained sand, (b) TCE-air in medium-grained sand, (c) water-air in fine-grained sand, and (d) TCE-air in fine-grained sand (from Fetter [1999]). Fetter, C.W., Contaminant Hydrogeology, 2nd Edition, 1999. Reprinted by permission of Pearson Education, Inc., Upper Saddle River, NJ.

models are the ones proposed by Brooks and Corey [1964] and van Genuchten [1980]. The Brooks and Corey model [1964] has the following form

$$S_e = \left(\frac{p_d}{p_c}\right)^{\lambda} \quad \text{for } p_c > p_d$$

$$S_e = 1 \quad \text{for } p_c > p_d \tag{2.6}$$

where λ is an index for the pore size distribution, p_d is the displacement pressure, θ_{wir} and S_{wir} are the irreducible wetting phase content and saturation, respectively, and the effective saturation is defined as

$$S_e = \frac{\theta_w - \theta_{wir}}{\phi - \theta_{wir}} = \frac{S_w - S_{wir}}{1 - S_{wir}} \tag{2.7}$$

The van Genuchten model [1980] has the following form

$$S_e = \left[1 + (\alpha h_c)^n\right]^{1-1/n} \tag{2.8}$$

where h_c is the capillary pressure head defined by Equation (2.5) and α and n are model parameters. From an overall perspective the two models may often describe the experimental data equally well. The major difference between the two models occurs near full wetting phase saturation. At this point, the Brooks and Corey model assumes a finite displacement pressure while the van Genuchten model provides a smooth transition (Figure 2.10). From physical reasoning, it is plausible to consider a finite displacement pressure at least for a porous material with a narrow pore size distribution. For more mixed porous media, the smooth transition may be more relevant. Also, the van Genuchten relation is more appropriate for field soils with a wide range of pores sizes, including large pores caused by worms and plants.

The parametric models of the capillary pressure curves were originally meant as a convenient way of representing the experimental values of capillary pressure and wetting fluid content. Subsequently, the models have been applied extensively and fitted to a large amount of experimental data (particularly to data for soil water retention curves). Carsel and Parrish [1988] estimated the water-air capillary pressure-saturation parameters for the Brooks and Corey and van Genuchten models for 12 major soil textural groups. Table 2.4 lists the average

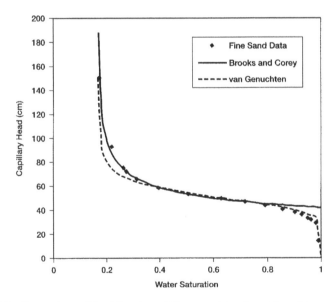

Figure 2.10. Comparison of fitted Brooks and Corey model and van Genuchten model (from Charbeneau [2000]). Charbeneau R.C., Groundwater hydraulics and pollutant transport, 1st Edition, 2000. Reprited by permission of Pearson Education, Inc., Upper Saddle River, NJ.

values for these soil classes, which may be useful to establish a first estimate of the capillary pressure-saturation characteristics if no specific measurements are available. The UNSODA database contains water-air, capillary pressure-saturation parameters for the Brooks and Corey and van Genuchten models for a large number of soils (Leij et al. [1996], http://www.ussl.ars.usda.gov/models/unsoda.HTM). A database developed by Beckett and Joy [2003, http://groundwater.api.org/lnapldatabase/] contains Brooks and Corey and van Genuchten model parameters for LNAPL-water systems in various soils.

Our discussion of capillary pressure and capillary pressure curves has concerned the case where only two phases are present in the porous medium. However, for NAPL spills in the unsaturated zone, three phases will coexist in the pores and consequently interfaces between several pairs of immiscible phases may occur. To understand the concepts of capillary pressure in a three-phase system, it is desirable to consider only two phases at a time and sets of two-phase capillary pressure curves. This approach works for a three-phase system if a clear wettability order for the phases can be assumed, as suggested by Parker [1989]. If we assume that the wettability order is water > NAPL > air, it implies that, at the pore scale, water is present in the pore space closest to the grain particles, NAPL occupies the pore space next to the water, and air occupies the remaining pore space in contact with NAPL (see Figure 2.5). On the continuum scale, the wettability assumption implies that the smaller pores are filled with water and the larger pores with air, while the intermediate sized pores are occupied primarily by NAPL.

This wettability order gives rise to the following simplifying assumptions regarding capillary pressure-saturation relationships [Parker, 1989] where we use p_{cij} to designate the capillary pressure between fluid phase pair ij ($ij = ao, aw, ow$).

TABLE 2.4. Average values of water-air capillary pressure-saturation parameters for 12 major soil textural groups [Carsel and Parrish, 1988].

Texture	ϕ	K_s (m/d)	θ_{wir}	α (cm^{-1})	n	h_d (cm)	λ
Sand	0.43	712.8	0.045	0.145	2.68	7	1.68
Loamy sand	0.41	350.2	0.057	0.124	2.28	8	1.28
Sandy loam	0.41	106.1	0.065	0.075	1.89	13	0.89
Loam	0.43	25.0	0.078	0.036	1.56	28	0.56
Silt	0.46	6.0	0.034	0.016	1.37	62	0.37
Silty loam	0.45	10.8	0.067	0.020	1.41	50	0.41
Sandy clay loam	0.39	31.4	0.100	0.059	1.48	17	0.48
Clay loam	0.41	6.2	0.095	0.019	1.31	53	0.31
Silty clay loam	0.43	1.7	0.089	0.010	1.23	100	0.23
Sandy clay	0.38	2.9	0.100	0.027	1.23	37	0.23
Silty clay	0.36	0.5	0.070	0.005	1.09	200	0.09
Clay	0.38	4.8	0.068	0.008	1.09	125	0.09

(a) The water content S_w is entirely determined by the capillary pressure between NAPL and water, i.e.

$$p_{cow} = p_{cow}(S_w) \qquad (2.9)$$

(b) The total liquid content $S_t = S_w + S_o$ (or alternatively the air saturation $S_a = 1 - S_t$) is entirely determined by the capillary pressure between air and NAPL, i.e.

$$p_{cao} = p_{cao}(S_t) \qquad (2.10)$$

From the definition of capillary pressure, it follows that the capillary pressure between air and water is not an independent quantity but is given as

$$p_{caw} = p_{cao} + p_{cow} \qquad (2.11)$$

As a consequence of these wettability assumptions, the local saturation distribution of a three-phase water-NAPL-air system is entirely determined by two-phase capillary pressure-saturation relationships for NAPL-water and NAPL-air.

The two-phase capillary pressure-saturation relationships required for estimating the phase distributions in a porous media may be measured directly. However, information on capillary pressure-saturation relationships for NAPL-water and NAPL-air pairs are often unavailable. Water retention characteristics (air-water capillary pressure curve) are more readily available for a given porous medium. In this case, the NAPL-water and air-NAPL relationships may be estimated using a scaling technique [Parker et al., 1987]. The capillary pressure-saturation relationship for a given fluid pair reflects the pore size distribution of the porous medium. For a given volumetric ratio between generic wetting and non-wetting phases, it may be assumed that the same geometry or the same radius of curvature between the two immiscible phases applies, regardless of the phases involved.

Lenhard and Parker [1987] inferred from Laplace's equation of capillarity (Equation (2.3)) that, for given porous medium and a given saturation ratio between two phases, the capillary pressure depends exclusively on the interfacial tension between the phases involved. Thus, we may transform a capillary pressure curve for one phase pair to a curve for another phase pair using the interfacial tension as a scaling parameter. Recalling that the air-water capillary pressure is most readily available, we can estimate the capillary pressure curves for NAPL-water and air-NAPL, respectively, for the same porous medium using the following equations

$$p_{cow}(S_w) = \frac{p_{caw}(S_w)}{\beta_{ow}}$$

$$p_{cao}(S_o) = \frac{p_{caw}(S_w)}{\beta_{ao}} \qquad (2.12)$$

where the scaling factors are defined as

$$\beta_{ow} = \frac{\sigma_{aw}}{\sigma_{ow}}$$

$$\beta_{ao} = \frac{\sigma_{aw}}{\sigma_{ao}}$$

(2.13)

and σ_{ij} is the interfacial tension for phase-pair (ij) [Lenhard and Parker, 1987].

Figure 2.11 demonstrates how scaling can be used to estimate capillary pressure curves for NAPL-water and air-NAPL, based on experimental data for air-water. The van Genuchten parametric model (Equation 2.8) was fitted to the air-water experimental data to obtain the van Genuchten parameters given in Table 2.5. The NAPL-water and air-NAPL curves were obtained with the same van Genuchten parameters and by applying the scaling factor parameters given in Table 2.5.

The results in Figure 2.11 demonstrate that the scaling approach is viable. However, the accuracy of the scaling is directly related to the accuracy of the interfacial tension estimates. According to Huntley and Beckett [2002b], field-measured values of fluid-pair interfacial tensions are often much smaller than laboratory-measured values. Since capillary pressure models (e.g. Brooks and Corey and van Genuchten) relate saturations to exponential functions of capillary pressures (see Equations 2.6 and 2.8), small errors in interfacial tensions can result in large errors in predicted saturations.

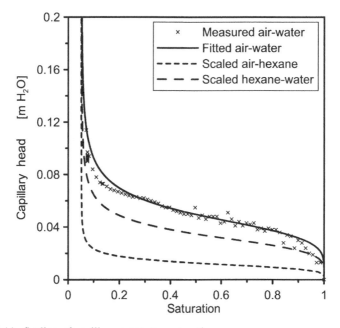

Figure 2.11. Scaling of capillary pressure – saturation curves.

TABLE 2.5.   Parametric values and scaling factors
applicable to the example shown in Figure 2.11.

van Genuchten parameters	
n	5.2
α	21.8 m^{-1}
S_{wir}	0.052
Interfacial tension (dyn/cm)	
σ_{aw}	72.8
σ_{ow}	51.0
σ_{ao}	18.4
Scaling factors	
β_{ow}	1.43
β_{ao}	3.95

2.2.3   Relative Permeability and Relative Permeability Curves

When two or more phases are simultaneously present in a porous medium, they will compete for the pore space. As a result, the permeability for each phase will be smaller than the intrinsic permeability when only that phase is present. This leads to the concept of relative permeability, which is defined as the ratio of the permeability of a phase at a given saturation to the intrinsic permeability

$$k_{r\beta} = \frac{k_\beta}{k}, \quad 0 \le k_\beta \le k \quad \text{and} \quad 0 \le k_{r\beta} \le 1 \qquad (2.14)$$

where k_β is the permeability for phase β at saturation S_β and k is the intrinsic permeability. The relative permeability varies as a function of saturation. The shape of the relative permeability function is a characteristic of the specific porous medium and depends on whether the phase is wetting or non-wetting with respect to the porous material.

Figure 2.12 shows typical relative permeability curves for wetting and non-wetting phases in a porous medium. Both curves begin at unity for full phase saturation and decrease to zero as the corresponding phase saturation decreases. The relative permeability of both phases approaches zero at saturations corresponding to the irreducible and residual saturations. Thus, the residual non-wetting phase saturation must be exceeded before the non-wetting phase will flow and, likewise, the irreducible wetting phase saturation has to be exceeded before the wetting phase will flow.

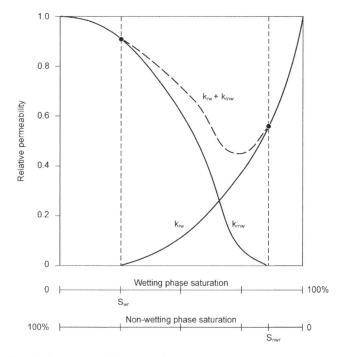

Figure 2.12. Relative permeability curves for a two-phase system.

According to the relative permeability curves in Figure 2.12, the maximum relative permeability for the wetting phase (at residual non-wetting-phase saturation) is less than the maximum value for the non-wetting phase (at irreducible wetting-phase saturation). This difference is caused by the fact that the non-wetting phase at residual saturation is found as isolated blobs in the largest pores and thus imposes an obstacle to the flow of the wetting phase. Whereas, the wetting phase at irreducible saturation is located in the finer pores, obstructing the flow of the non-wetting phase to a lesser degree. We also see in Figure 2.12 that, for comparable phase saturations, $k_{rnw} > k_{rw}$. This characteristic is attributed to wettability effects: the wetting phase surrounds the porous media skeleton and thus, due to its lubrication effect, reduces the resistance to flow for the non-wetting phase. Finally, we note from Figure 2.12 that when both phases are present, $k_{rnw} + k_{rw} < 1$. This result reflects the fact that the flow of each fluid phase interferes with the other's flow, due to differences in fluid viscosities.

Relative permeability functions can be determined in the laboratory on porous medium cores using various direct and indirect techniques (see Sale [2001] for a discussion of various methods). Direct techniques involve imposing steady-state flows of both phases on the core. By measuring the saturations and the pressure differences in the phases across the core, the hydraulic conductivities of each phase can be determined by applying Darcy's law. By varying the flow rates of the phases, individual points on the relative permeability curves can be determined. Indirect methods involve performing dynamic displacement experiments while measuring responses of relevant variables with time at various locations within the core. Both direct and indirect methods are, in general, elaborate and tedious to perform.

To circumvent these difficulties, predictive models for relative permeability have been proposed that use the information from the more easily measured capillary pressure-saturation data. These models are based on the idea that the capillary pressure curves possess information on the pore size distribution of the media. Burdine [1953] and Mualem [1976] have derived well-known models of this form. These models are based on functional forms of the capillary pressure curves, such as the van Genuchten or Brooks-Corey parametric models (Equations 2.6 and 2.8). For example, combining the van Genuchten capillary pressure-saturation parametric model (Equation 2.8), with Mualem's [1976] model, one obtains [Parker et al., 1987]

$$k_{rw} = S_{ew}^{1/2}[1-(1-S_{ew}^{1/m})^m]^2$$
$$k_{rnw} = (1-S_{ew})^{1/2}(1-S_{ew}^{1/m})^{2m}$$

$$(2.15)$$

where $m = 1 - 1/n$ and S_{ew} denotes the effective saturation defined in Equation (2.7).

As an example, Figure 2.13 shows relative permeability curves predicted from Equation (2.15) and based on the water-air capillary pressure curve shown in Figure 2.11 and associated parametric values listed in Table 2.5.

When three phases are present in the porous medium at the same time, measurement and prediction of relative permeability functions becomes considerably more complicated. However, if we can invoke the wettability assumptions mentioned earlier (wettability order: water > NAPL > air), we can assume that the relative permeability of water is entirely a function of the water saturation and that the relative permeability of air is entirely a function of the air saturation. Since NAPL is the phase of intermediate wettability, its relative permeability is not only a function of the NAPL saturation, but also depends on the saturation of water or air [Corey, 1994]. Parker et al. [1987] extended the two-phase expressions (Equation 2.15) to three phases and arrived at the following parametric models

$$k_{rw} = S_{ew}^{1/2}[1-(1-S_{ew}^{1/m})^m]^2$$
$$k_{ra} = (1-S_{et})^{1/2}(1-S_{et}^{1/m})^{2m}$$
$$k_{ro} = (S_{et}-S_{ew})^{1/2}[(1-S_{ew}^{1/m})^m - (1-S_{et}^{1/m})^m]^2$$

$$(2.16)$$

where the total effective liquid saturation, S_{et}, is defined as

$$S_{et} = \frac{S_w + S_o - S_{wir}}{1 - S_{wir}}$$

$$(2.17)$$

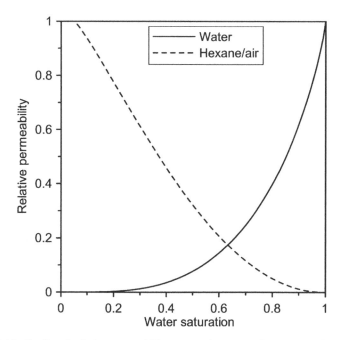

Figure 2.13. Predicted relative permeability curves for a two-phase system based on the capillary pressure curve shown in Figure 2.11.

Other forms of the NAPL relative permeability have also been proposed; see Corey [1994] and Helmig [1997] for a review of various relations.

2.2.4 Darcy's Law and Governing Equations for Multiphase Flow

Flow of a fluid in porous media is determined by the interplay of pressure, gravitational, viscous, and inertial forces. The pressure and gravitational forces constitute the driving forces, whereas viscous forces constitute the resisting force. Inertial forces can, for all practical purposes, be neglected for flow in porous media. The basic equation for fluid flow in porous media, Darcy's law, is based on the balance between pressure, gravitational, and viscous forces. For flow in a porous medium fully saturated with water, Darcy's law is

$$\bar{q} = -K\nabla h = -\frac{\rho_w g k}{\mu_w}\nabla h \qquad (2.18)$$

where $\bar{q}$ is the Darcy velocity vector, K is the hydraulic conductivity representing the viscous resistance to flow, and g is gravitational acceleration. The hydraulic conductivity, K, can be decomposed into properties representing fluid characteristics (density ρ_w and viscosity μ_w) and porous medium characteristics (intrinsic permeability k). We note that Darcy's law is based on the assumption that viscous forces are linearly proportional to the flow rate.

The hydraulic head is defined as

$$h = \frac{p}{\rho_w g} + z \tag{2.19}$$

where p is water pressure and z is the upward vertical coordinate. Substituting the definition of hydraulic head (Equation 2.19) into Darcy's law (Equation 2.18) yields

$$\overline{\mathbf{q}} = -\frac{k}{\mu}(\nabla p + \rho g \nabla z) \tag{2.20}$$

where ∇z is the upward unit vector and k is the intrinsic permeability.

Darcy's law has been extended to multiphase systems based on the assumptions that (a) the driving force for a phase β is determined by the gradient in the phase pressure and the gravitational force, and (b) the relative permeability of the phase, $k_{r\beta}$, is a function of the saturations of the phases involved. The generalized form of Darcy's law can be expressed in the following form [Aziz and Settari, 1986]

$$\overline{\mathbf{q}}_\beta = -\frac{k_{r\beta}k}{\mu_\beta}(\nabla p_\beta + \rho_\beta g \nabla z) \tag{2.21}$$

Assuming that the porous skeleton is incompressible, the mass conservation equation for phase β is

$$\phi\frac{\partial \rho_\beta S_\beta}{\partial t} = -\nabla \cdot (\rho_\beta \overline{\mathbf{q}}_\beta) \tag{2.22}$$

Substituting Darcy's law (Equation 2.18) into Equation (2.22) yields a partial differential equation (PDE) for phase β

$$\phi\frac{\partial \rho_\beta S_\beta}{\partial t} = -\nabla\left[\rho_\beta \frac{k_{r\beta}k}{\mu}(\nabla p_\beta + \rho_\beta g \nabla z)\right] \tag{2.23}$$

For a porous medium in which three phases are present, a PDE can be developed for each phase. The three equations are coupled since both relative permeability and pressures are functions of saturations. Assuming that the basic fluid characteristics (density and viscosity) and porous medium characteristics (porosity, intrinsic permeability and relative permeability) are known, six unknowns remain: saturation and pressure for each of the phases. To solve the set of coupled PDEs, three additional equations are required.

The first equation is obtained from the total volume balance for the phases:

$$S_w + S_o + S_a = 1 \tag{2.24}$$

The second and third equations come from the capillary pressure-saturation Equations (2.9) and (2.10):

$$\begin{aligned} p_{cow}(S_w) &= p_o - p_w \\ p_{cao}(S_t) &= p_a - p_o \end{aligned} \tag{2.25}$$

As mentioned earlier, the capillary pressure between air and water is not independent and is determined by the other two capillary pressures as in equation (2.11).

The system of PDEs for the simultaneous flow of three immiscible phases water, NAPL and air (Equation 2.23) can now be coupled by introducing the saturation condition (Equation 2.24) and the two capillary pressure relations (Equation 2.25) and by specifying fluid and medium properties including relative permeability functions. Since the equations are highly non-linear, advanced numerical simulation techniques are required to solve the problem and a significant computational effort can be expected, particularly for three-dimensional problems [Helmig, 1997].

From a practical point of view, simplifications that reduce the dimensionality of the problem and the number of equations are desirable. For example, a NAPL spill in the unsaturated zone may be treated as a one-dimensional, single-phase flow problem if the following assumptions are adopted. First, it may be a viable approximation to treat the water phase as if it were at irreducible saturation. The intruding NAPL is thus flowing on the stationary water phase while displacing air. Second, it is common to assume that the air phase in the unsaturated zone is at atmospheric pressure throughout the domain and does not impose a resistance to the infiltration of NAPL. In many cases, this is a reasonable assumption because both the density and viscosity of air are much lower than those for liquids. Consequently, the air's high mobility will rapidly eliminate gradients in the air pressure as long as the air phase remains in contact with the atmosphere. However, the presence of the air phase will impact both the relative permeability and the NAPL pressure.

Thus, on the basis of the above assumptions, only the PDE for NAPL needs to be solved. The PDE is analogous to the Richards' equation for unsaturated flow (see, e.g., Jury et al. [1983]). Furthermore, if the spill is of sufficient lateral extent, lateral spreading due to capillary forces is negligible compared to the size of the spill. The migration is thus reduced to a 1-D flow problem and relatively simple solutions may be obtained if certain assumptions are made with respect to the NAPL saturation profile and the relative permeability of NAPL.

These approximations lead to relativly simple models for NAPL infiltration and redistribution in the unsaturated zone that may capture the important features of the processes. Reible et al. [1990] and Weaver [1994], among others, have assumed that the NAPL profile has the shape of a plug (uniform saturation behind a sharp front) during infiltration into a homogeneous porous medium. These authors developed a NAPL flow model similar to the Green and Ampt [1911] model for water infiltration. Reible et al. [1990] assumed that, during redistribution (after the NAPL source has been exhausted), both the wetting and drainage fronts are sharp. Weaver et al. [1994] employed the kinematic wave approach for modeling the development of the drainage process.

2.3   NAPL MASS TRANSFER AND TRANSPORT

Summary: Chemicals in the subsurface zone undergo phase partitioning between the gas, aqueous, NAPL, and solid phases. The equilibrium partitioning of a chemical into different phases means that only a fraction of the total chemical mass will be present in any single phase. Vapor pressure and solubility are parameters that determine the amount of partitioning between gas and liquid phases and between the aqueous phase and NAPL. Equilibrium compositions of gas and liquid phases containing mixtures of components may be approximated using Raoult's Law. Adsorption onto the solid phase occurs primarily from the aqueous phase, rather than from the NAPL or gas phases. The adsorbed concentration is sometimes a linear function of the aqueous concentration, especially at low aqueous concentrations. If chemical transport occurs in one phase, the rate of transport is retarded by the overall partitioning into the other phases, relative to the rate of transport that would occur without partitioning. In addition to equilibrium partitioning, the fate of NAPL-derived chemicals can migrate in the gas, aqueous, and NAPL phases, via advection, molecular diffusion, and mechanical dispersion.

At sites contaminated with NAPLs, there are as many as three fluid phases present in the pore space at any location. These fluid phases are the gas phase, the aqueous phase, and the NAPL, and they are considered to have distinct phase boundaries with each other. Each fluid phase may be composed of several components or compounds. The gas phase contains a mixture of gases such as nitrogen, oxygen, carbon dioxide, water vapor, and in contaminated locations, organic chemical vapors. The aqueous phase is mostly liquid water, but contains dissolved organic and inorganic solutes, as well as dissolved gases. The NAPL phase may be a pure chemical such as trichloroethylene (TCE), or it may be a mixture of several organic compounds such as gasoline, which is composed of hundreds of hydrocarbons. When considering the movement of chemicals in the subsurface, it is very important to distinguish between phases and components. For example, it is incorrect to call TCE a NAPL unless it is present as a separate

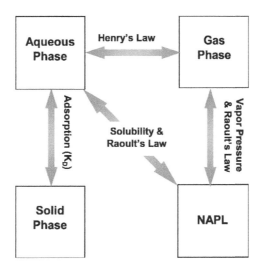

Figure 2.14. Conceptual view of phase equilibrium relationships used for interphase mass transfer of chemical components.

nonaqueous liquid phase. If the TCE is dissolved in water, or if it exists as a vapor in the gas phase, it is not NAPL.

Chemicals in the subsurface zone undergo phase partitioning between the gas, aqueous, NAPL, and solid phases, as described schematically in Figure 2.14. This process is transient and is known as kinetic interphase mass transfer. The driving force for any type of interphase mass transfer is the degree of chemical potential disequilibrium between the phases, and the final state is one of chemical potential equilibrium between the phases. In many cases, the rate of kinetic interphase mass transfer is fast enough to allow the assumption of local chemical equilibrium.

2.3.1   Vapor Pressure and Solubility

Two very important properties of any pure chemical are its vapor pressure (p_{vap}^i), and its aqueous solubility $(\bar{C}_w^i)$. The vapor pressure of a compound is the gas partial pressure of the compound in equilibrium with its pure liquid or solid form. A chemical's vapor pressure is a strong function of temperature, as shown in Figure 2.15. The curve in Figure 2.15 shows the maximum (saturated vapor) value of the gas partial pressure (p_a^i) as a function of temperature. Because this is the maximum possible partial pressure for the component, if the system conditions are changed, it is possible for a vapor to condense into a liquid phase. This phenomenon occurs, for example, when a saturated vapor is cooled. If the saturated vapor is an organic chemical such as TCE, it will condense into a NAPL when cooled.

The vapor pressure also determines rates of evaporation of a liquid. If the partial pressure of a chemical is below the vapor pressure line, then it is not in

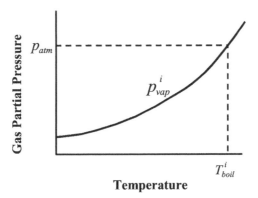

Figure 2.15. Chemical saturated vapor pressure as a function of temperature. A liquid boils when its vapor pressure is equal to the total pressure (atmospheric in this case).

equilibrium with the liquid form of the chemical. This state of chemical disequilibrium thus drives the evaporation process, until the gas partial pressure reaches the vapor pressure, or the liquid phase is depleted. In this way, a NAPL phase may evaporate, and eventually disappear if clean gas flows through it.

The aqueous solubility of a compound is the dissolved aqueous concentration of the compound in equilibrium with its pure liquid or solid form (or with a specified gas partial pressure). A chemical's solubility is generally a weak function of temperature. The solubility is, however, very sensitive to certain classes of dissolved chemicals called surfactants and cosolvents. Surfactants and cosolvents have the property of greatly increasing the solubility of hydrophobic (low solubility) organic compounds in aqueous solutions.

If the concentration of a chemical in water begins to exceed its solubility, a separate solid or liquid phase can precipitate from the solution. In the case of dissolved organic chemicals, that new phase would be a NAPL. The aqueous solubility also determines the rates of dissolution for a NAPL. If the aqueous concentration of a chemical that forms a pure NAPL is below the solubility, then it is not in equilibrium with the NAPL. This state of chemical disequilibrium drives the dissolution process, until the aqueous concentration reaches the aqueous solubility, or the NAPL is depleted. Thus the NAPL phase may dissolve, and eventually disappear, if water at concentrations less than solubility flows through the NAPL-contaminated region.

2.3.2 Phase Densities

Knowledge of units and methods for calculating phase densities and phase concentrations is important for understanding the distribution of chemicals in the subsurface. This topic can be confusing due to the different conventions, units, and nomenclature in common use in the literature. In the following, we establish conventions that will be followed in the remainder of the text.

The water, gas, and NAPL phase densities, ρ_w, ρ_a, and ρ_n are functions of phase composition, temperature, and pressure. The density of multicomponent liquids (NAPL and aqueous phases) may be computed to a good approximation (typically within 1%) by assuming volume additivity. That is, the volume of a mixture of components is assumed to be equal to the sum of the individual component volumes. Given the pure component densities of N components, the phase density is calculated as a function of composition using a volume-weighted average of the individual component densities:

$$\rho_{mix} = \frac{\sum\limits_{i=1}^{N} \rho_i V_i}{\sum\limits_{i=1}^{N} V_i} \tag{2.26}$$

where ρ_i and V_i are the density and volumes of the ith component, respectively [Mercer and Cohen, 1990].

The aqueous and NAPL phases are only slightly compressible, and in most environmental problems, liquid compression effects are expected to be negligible. The liquid phase densities are weak functions of temperature. For example, pure water has a density of 999.7 kg/m^3 at 10°C, and this drops to a value of 958.4 kg/m^3 at a temperature of 100°C [Bejan, 1984]. NAPL phase densities also are relatively insensitive to temperatures in the range commonly encountered in the subsurface. However, since NAPL phase densities are strong functions of composition, calculation of phase densities require component fractions and individual densities in the case of multicomponent NAPLs.

The gas phase density is very sensitive to variations in composition, temperature, and pressure. The density of the gas phase is calculated using the real gas law:

$$\rho_a = \frac{p_a M_{wt}}{ZRT} \tag{2.27}$$

where p_a is the total gas phase pressure, M_{wt} is the average molecular weight of the gas phase [kg/mol], Z is the gas compressibility factor, R is the universal gas constant [8.3144 N m/mol K], and T is the absolute temperature [K]. The assumption of ideal gas behavior is often appropriate for environmental studies, due to the relatively small variations in gas temperature and pressure. In this case, the gas compressibility factor is equal to one. Following Dalton's Law (see, e.g., Sontag and Van Wylen [1982]), the ideal gas law can be written as

$$\rho_a = \frac{\sum\limits_{i=1}^{N} p_a^i M_{wt}^i}{RT} \tag{2.28}$$

where p_a^i is the gas partial pressure of component i, M_{wt}^i is the molecular weight of component i, and N is the total number of gas phase components. The strong

influence of composition, temperature, and pressure on gas phase density often
gives rise to buoyancy driven gas flows in the vadose zone. Comparing the density of a gas computed using Equation (2.28) with that of air ($M_{wt}^{air} \approx 29$ g/mol),
it is apparent that if a gas has an average molecular weight greater than that of air,
then the gas will be denser than air. The reverse is also true; low molecular weight
gases such as methane are lighter than air.

---

*Example: Calculate the gas density of a mixture of trichloroethylene (TCE)
vapor and air at one atmosphere total pressure (101,325 Pa) if the temperature is 20°C and the TCE partial pressure is equal to its pure vapor pressure
of 7,800 Pa. The molecular weight of TCE is 131.4 g/mol. How does this
mixture density compare with the density of pure air?*

*Solution: Given that the total pressure was specified, the air partial pressure
is 101,325 – 7,800 = 93,525 Pa. Then the gas density using equation (2.28) is:*

$$\rho_a = \frac{(7800)(0.1314) + (93525)(0.029)}{8.3144(293.15)} = 1.53 \, kg/m^3$$

*This is considerably higher than the value of 1.21 kg/m³ for pure air at the
same temperature.*

2.3.3   Concentrations

The fact that a given chemical or compound may be present in the gas, aqueous, or NAPL phases, combined with the multidisciplinary nature of environmental research, has resulted in common usage of several different, but related,
measures of chemical concentration.

The mass concentration of a chemical, C_β^i, is the mass of chemical i per unit
volume of the phase β, where β can be the gas, aqueous, NAPL, or solid (for
chemical adsorption) phase. The mass concentration has units of kg/m³, and is
often reported as mg/l or μg/l. Concentrations of chemical components in aqueous phase samples are typically reported as mass concentrations. When the
density of the aqueous phase is ~1000 g/l, one mg/l is approximately 1 part per
million (ppm, a mass fraction).

For fluid phases (gas, aqueous, NAPL), the phase density is equal to the sum
of all of the chemical concentrations in a phase:

$$\rho_\beta = \sum_{i=1}^{N} C_\beta^i \tag{2.29}$$

A total mass concentration, C_T^i, can also be defined on a total volume basis.
The total concentration of a chemical includes the sum of the amounts of

the chemical in each of the fluid phases, as well as the adsorbed concentration [Jury et al., 1983; Feenstra et al., 1991; Mott, 1995; Mariner et al., 1997]:

$$C_T^i = \phi S_a C_a^i + \phi S_w C_w^i + \phi S_o C_o^i + C_s^i = \left[\frac{m^i}{V_T}\right] \tag{2.30}$$

where C_s^i is the adsorbed mass of chemical i per unit bulk volume of the porous medium.

The mass fraction of a chemical, X_β^i, is the mass of the chemical i per unit mass of the phase β. The sum of the component mass fractions in a phase is equal to one, and the chemical mass fraction in a phase is equal to the mass concentration divided by the phase density

$$X_\beta^i = \frac{C_\beta^i}{\rho_\beta} = \left[\frac{m_\beta^i}{m_\beta}\right] \tag{2.31}$$

Mass fractions are dimensionless and are often reported as weight-percent or parts per million. We can also define the total mass fraction of a chemical insoil as the mass of chemical extant to all phases per unit mass of dry (clean) porous medium:

$$X_T^i = \frac{C_T^i}{\rho_b} = \left[\frac{m_T^i}{m_s}\right] \tag{2.32}$$

where ρ_b is the dry bulk density of the porous medium. The total mass fraction is commonly used to describe chemical concentrations in soil borings and rock samples.

Molar concentration (c_β^i) is similar to mass concentration, and is defined as the number of moles of component i per unit volume of phase β. It is mainly used for aqueous concentrations, where units are mol/m^3 or mmol/l. The sum of molar concentrations of all components in a phase gives the phase molar density:

$$c_\beta = \sum_{i=1}^{N} c_\beta^i \tag{2.33}$$

The mole fraction, χ_β^i, is used in many chemical calculations and is particularly convenient for problems involving ideal gases. The mole fraction of a chemical is defined as the number of moles of component i per mole of the β-phase, so it is dimensionless like the mass fraction. The sum of the mole fractions of all components in a phase is equal to one, and the mole fraction is equal to the molar concentration divided by the molar density. The mole fraction is related to the mass fraction through the component molecular weights:

$$\chi_\beta^i = \frac{X_\beta^i / M_{wt}^i}{\sum_{i=1}^{N} X_\beta^i / M_{wt}^i} \tag{2.34}$$

and

$$X_\beta^i = \frac{\chi_\beta^i M_{wt}^i}{\sum_{i=1}^{N} \chi_\beta^i M_{wt}^i} \tag{2.35}$$

Gas-phase concentrations are measured and reported in a variety of related units. Assuming ideal gas behavior, the gas mass concentration may be computed from the gas partial pressure:

$$C_a^i = \frac{p_a^i M_{wt}^i}{RT} \tag{2.36}$$

where p_a^i is the partial pressure of component i. The gas mole fraction is determined by the ratio of the partial pressure to the total pressure

$$\chi_a^i = \frac{p_a^i}{p_a} \tag{2.37}$$

where p_a is the total gas phase pressure. The gas volume fractions are also equal to gas mole fractions for ideal gases, and a common unit for the measurement of gas concentrations is the part per million by volume (ppmv). This unit is completely different from a mass fraction ppm used to describe aqueous mass fractions.

Example: Assuming ideal gas behavior, calculate the gas phase mass concentration, mole fraction, and volume fraction corresponding to a TCE partial pressure of 500 Pa, given a temperature of 20°C, and a total gas pressure of 100,000 Pa.

Solution: Using Equation (2.36), the gas phase mass concentration is

$$C_a^{TCE} = \frac{500(0.1314)}{8.3144(293.15)} = 0.027 \ kg/m^3 \ or \ 27mg/l$$

Using Equation 2.37, the mole fraction is

$$\chi_a^{TCE} = \frac{500}{100000} = 0.005$$

The gas volume fraction is equal to the mole fraction, and it has a value of 5000 in units of ppmv.

2.3.4 Equilibrium Phase Partitioning

Phase equilibrium between a gas phase and a NAPL depends on the chemical makeup of the NAPL. If the NAPL consists mainly of a single chemical, then the equilibrium gas partial pressure (p_a^i) is equal to the chemical vapor pressure, p_{vap}^i. Then the gas concentration is given by Equation (2.36) or (2.37). If the NAPL is composed of many compounds, the multicomponent chemical equilibrium is more complex and is, in general, a function of the molar composition [Prausnitz et al., 1986]. To a first approximation, the equilibrium gas composition may be calculated using Raoult's Law, where the gas partial pressure is a linear function of the chemical mole fraction in the NAPL (assuming activity coefficients are equal to unity):

$$p_a^i = \chi_o^i p_{vap}^i \tag{2.38}$$

Thus, the equilibrium partial pressure of a chemical in equilibrium with a multi-component NAPL is always less than the chemical's pure vapor pressure.

Phase equilibrium between the gas phase and the aqueous phase often involves relatively dilute aqueous concentrations. In the limit of dilute solution (a relative definition), equilibrium gas and aqueous concentrations have a linear relationship, known as Henry's law. Henry's constant, H^i, is generally defined as the ratio of the gas concentration to the aqueous concentration. A variety of different units are used for the gas and aqueous concentrations. Some common definitions of Henry's constant include:

$$H_\chi^i = \frac{p_a^i}{\chi_w^i} \tag{2.39}$$

where H_χ^i has units of pressure,

$$H_m^i = \frac{p_a^i}{c_w^i} \tag{2.40}$$

where H_m^i has units of pressure × volume / mole, and

$$H_C^i = \frac{C_a^i}{C_w^i} \tag{2.41}$$

in which H_C^i is dimensionless.

The value of Henry's constant varies widely, depending on the chemical. For example, at room temperature, n-octane has a dimensionless Henry's constant (H_C^i) equal to 140.0, while phenol has a dimensionless Henry's constant of 7×10^{-6} [Jury et al., 1984]. Many commonly encountered aromatic and halogenated aliphatic organic compounds with moderate solubility have dimensionless Henry's constants in the range of 0.1 to 1.0, indicating that these compounds will have a tendency to volatilize from the aqueous phase. Henry's constant is a strong function of temperature, where it increases with increasing temperature.

Example: Calculate the aqueous mass concentration of TCE in equilibrium with a gas phase TCE volume fraction of 400 ppmv at a temperature of 20°C and a total pressure of 100,000 Pa. The vapor pressure of TCE at this temperature is 7800 Pa, and the aqueous solubility of TCE is 1100 mg/l.

Solution: Using Equation (2.36), the TCE vapor concentration corresponding to the vapor pressure is calculated to be

$$C_a^{TCE} = \frac{7800(0.1314)}{8.3144(293.15)} = 0.42 \; kg/m^3, \; or \; 420 \; mg/l.$$

Assuming that Henry's law is valid up to the TCE solubility limit, H_C^{TCE} can be estimated as the ratio of the saturated TCE vapor concentration to the TCE aqueous solubility, using Equation (2.41):

$$H_C^{TCE} = \frac{420}{1100} = 0.38$$

The TCE volume fraction of 400 ppmv corresponds to a mole fraction of 0.0004. Using Equation (2.38) with a total pressure of 100,000 Pa gives the TCE partial pressure of 40 Pa. This value is used in Equation (2.36), to get the TCE gas mass concentration of 0.0022 kg/m³ or 2.2 mg/l. Finally, the dimensionless Henry's law (Equation 2.41) is rearranged to give

$$C_w^{TCE} = \frac{2.2}{0.38} = 5.8 \; mg/l$$

Note that this example may also be solved by considering the ratios of the chemical concentrations to their saturated values. This technique is discussed later in this section.

Equilibrium between the NAPL and the aqueous phases may be calculated by first computing the equilibrium gas phase concentration and then applying Henry's law. For single component NAPLs (with a limited aqueous solubility), Equations (2.36) through (2.38) can be used with the chemical's vapor pressure to compute the NAPL solubility. If the NAPL is multicomponent, then Raoult's Law is used to determine the gas equilibrium of each component, followed by

Henry's Law to get the aqueous equilibrium for each component. This method is equivalent to weighting the pure aqueous solubility of the individual compounds by their mole fraction in the NAPL [Banerjee, 1984; Cline et al., 1991; Lee et al., 1992]:

$$C_w^i = \chi_o^i \overline{C}_w^i \tag{2.42}$$

Here, $\overline{C}_w^i$ is the aqueous solubility of a pure compound i and we are assuming activity coefficients equal to unity. In this multicomponent NAPL case, the overall aqueous NAPL solubility is the sum of the individual component solubilities calculated by Equation (2.42).

Many organic, inorganic, and radioactive contaminants partition strongly to the solid soil or rock grains. This phenomenon is known as adsorption, and it results from a variety of physical, chemical, and biological mechanisms. Because water is usually the wetting phase in the subsurface, the soil or rock grains are almost always covered with a layer of liquid water, even in systems with a thick vadose zone. Therefore, adsorption onto the solid phase occurs primarily from the aqueous phase, rather than from the NAPL. The relationship between adsorbed and aqueous concentration at equilibrium is determined through a series of experiments where uncontaminated solid phase is brought into contact with aqueous phase at varying concentrations. The results of these experiments can be plotted as a function of aqueous concentration to produce an adsorption isotherm. The isotherm data or a function fitted to the data provides a unique relationship between adsorbed and aqueous concentration at equilibrium.

The adsorbed concentration is sometimes a linear function of the aqueous concentration, especially at low aqueous concentrations. Linear adsorption isotherms are characterized by K_d^i, the soil-water distribution coefficient for a chemical. This coefficient is the ratio of the adsorbed mass fraction to the aqueous mass concentration:

$$K_d^i = \frac{X_s^i}{C_w^i} \tag{2.43}$$

and it has units of inverse concentration, m^3/kg. The solid adsorbed mass fraction is related to the adsorbed mass concentration through the dry bulk density of the soil or rock:

$$C_s^i = \rho_b X_s^i = \rho_b K_d^i C_w^i \tag{2.44}$$

The distribution coefficient is a function of both the chemical and the solid, and the value of K_d^i ranges over several orders of magnitude for various chemicals and solids [Fetter, 1999].

The equilibrium partitioning of a chemical into different phases means that only a fraction of the total chemical mass will be present in any single phase.

If chemical transport occurs in one phase, the rate of transport is retarded by the overall partitioning into the other phases, relative to the rate of transport that would occur without partitioning. An equilibrium multiphase retardation coefficient can be defined as the ratio of the total concentration of a chemical to its mass concentration in a single fluid phase (per unit total volume):

$$R^i_\beta = \frac{C^i_T}{\phi S_\beta C^i_\beta}$$

(2.45)

The concept of chemical retardation is only valid when there is a linear relationship between the total concentration (C^i_T), and the individual phase concentrations (C^i_a or C^i_w).

We can derive retardation coefficients for two general cases: first, where no NAPL is present and, second, where NAPL is present. For the first case, if the chemical is transported by the gas phase, its movement is slowed by a factor of R^i_a,

$$R^i_a = 1 + \frac{S_w}{S_a H^i_C} + \frac{\rho_b K^i_d}{\phi S_a H^i_C}$$

(2.46)

Similarly, if the chemical is transported by the aqueous phase, its movement is slowed by a factor of R^i_w,

$$R^i_w = 1 + \frac{S_a H^i_C}{S_w} + \frac{\rho_b K^i_d}{\phi S_w}$$

(2.47)

For the second case, when a single component NAPL is present, the equilibrium gas and aqueous concentrations are fixed at their saturation values (vapor pressure and solubility), and the total concentration depends mainly on the NAPL saturation. The retardation coefficient cannot be used under these conditions because the total concentration is not a linear function of the phase concentrations.

However, if a chemical exists only as a "dilute" component of a NAPL, we can calculate retardation coefficients in the following manner. We define a partition coefficient similar to Henry's Constant:

$$K^i_p = \frac{C^i_o}{C^i_w}$$

(2.48)

Then the gas and aqueous phase retardation coefficients include a term for the chemical partitioning in the bulk NAPL:

$$R^i_a = 1 + \frac{S_w}{S_a H^i_C} + \frac{S_o K^i_p}{S_a H^i_C} + \frac{\rho_b K^i_d}{\phi S_a H^i_C}$$

(2.49)

and

$$R_w^i = 1 + \frac{S_a H_c^i}{S_w} + \frac{S_o K_p^i}{S_w} + \frac{\rho_b K_d^i}{\phi S_w} \qquad (2.50)$$

Equations (2.49) and (2.50) may be used to analyze the results of NAPL partitioning interwell gas tracer tests conducted in the vadose zone and saturated zone, respectively (see, e.g., Jin et al. [1995]).

The concept of a total chemical concentration, given by Equation (2.30), can be combined with the equilibrium phase partitioning relationships to easily analyze multiphase partitioning problems involving a single component NAPL. We begin the analysis by determining the total concentration in the absence of NAPL. Assuming linear partitioning, the total concentration of a chemical is a linear function of either the aqueous or the gas concentration. In the absence of NAPL, Equation (2.30) can be combined with Equations (2.41) and (2.44) to obtain

$$C_T^i = \left(\phi S_a + \phi \frac{S_w}{H_C^i} + \rho_b \frac{K_d^i}{H_C^i} \right) C_a^i \qquad (2.51)$$

and

$$C_T^i = (\phi S_a H_C^i + \phi S_w + \rho_b K_d^i) C_w^i \qquad (2.52)$$

The maximum possible value for C_T^i in the absence of NAPL, $\bar{C}_T^i$, is computed using Equation (2.51) or (2.52) with the gas phase concentration calculated using the chemical vapor pressure, or with the aqueous solubility, respectively.

If NAPL is present in the sample, the NAPL saturation may be calculated from the definition of the total concentration (remembering that the gas and aqueous concentrations will be at their respective maxima at equilibrium):

$$S_o = \frac{C_T^i - \bar{C}_T^i}{\phi \rho_o} \qquad (2.53)$$

Here, the chemical concentration in the NAPL phase has been replaced by the NAPL density, since we are considering a single component NAPL. Equation (2.53) neglects the small loss of gas or water volume taken up by the NAPL phase, but this is insignificant in real calculations.

The concept of the total concentration may be used to determine whether or not NAPL is present in a soil sample (see, e.g., Feenstra et al. [1991]). First, the amount of contamination in the soil sample (which is typically reported as a mass fraction based on the dry density of the soil) is converted to a total concentration (C_T^i) using Equation (2.32). This total concentration is then compared to the theoretical maximum in the absence of NAPL, $\bar{C}_T^i$. If the actual total concentration is much greater than the theoretical maximum without NAPL (i.e. $C_T^i \gg \bar{C}_T^i$), then NAPL is clearly present in the sample. Similarly, if $C_T^i \ll \bar{C}_T^i$, then NAPL is not present in the sample. However, we note that the apparent absence of NAPLs in soil samples may simply be due to the volatilization of the NAPL as the sample is exposed to the atmosphere, given the volatile nature of many NAPLs.

Example: Suppose that a soil sample from a hazardous waste site is analyzed, and found to have a total carbon tetrachloride mass fraction of 15,300 ppm (mg/kg). Given the following soil properties: $\phi = 0.35$, $\rho_b = 1600\ kg/m^3$, $S_w = 0.52$, *and* $T = 25°C$ *and chemical properties:* $p_{vap}^{c-tet} = 15,100\ Pa$ $M_{wt}^{c-tet} = 153.8\ g/mole$, $H_C^{c-tet} = 1.24$, $K_d^{c-tet} = 0.0022\ m^3/kg$ *and* $\rho_o = 1584\ kg/m^3$, *determine the NAPL saturation in the sample, and the equilibrium gas and aqueous phase mass concentrations.*

Solution: The first step is to convert the measured carbon tetrachloride mass fraction into a total concentration using Equation (2.32):

$$C_T^{c-tet} = 0.0153(1600) = 24.48\ kg/m^3$$

Next, the saturated vapor concentration is computed from Equation (2.36) to be 0.937 kg/m³. Using the dimensionless Henry's law, Equation (2.41) gives an aqueous solubility of 0.755 kg/m³. These maximum phase concentrations are then used in either Equation (2.51) or (2.52) to get the $\bar{C}_T^{c-tet}$ *for this case. Using Equation (2.52),*

$$\bar{C}_T^{c-tet} = \{(0.35)(0.48)(1.24) + (0.35)(0.52) + (1600)(0.0022)\}(0.755) = 2.95\ kg/m^3$$

Clearly, $C_T^{c-tet} \gg \bar{C}_T^{c-tet}$ *so NAPL is present in the sample. Using Equation (2.53), the NAPL saturation is*

$$S_o = \frac{24.48 - 2.95}{(0.35)(1584)} = 0.039$$

Because NAPL is present, the gas concentration is equal to the saturated value of 0. 937 kg/m³ (937 mg/l) and the aqueous concentration is equal to the solubility of 0.755 kg/m³ (755 mg/l).

If NAPL is not present in the sample, then the equilibrium gas, aqueous, or adsorbed concentrations may be quickly computed from the total concentration (or vice-versa) using the following ratios:

$$\frac{C_a^i}{\bar{C}_a^i} = \frac{C_w^i}{\bar{C}_w^i} = \frac{C_s^i}{\bar{C}_s^i} = \frac{C_T^i}{\bar{C}_T^i} \tag{2.54}$$

where C_s^i is the adsorbed concentration.

> *Example: Suppose that another soil sample from the same hazardous waste site as before is analyzed, and found to have a total carbon tetrachloride mass fraction (X_T) of 15 ppm (mg/kg). Determine whether or not NAPL is present in the sample, and calculate the equilibrium gas and aqueous phase concentrations and the NAPL saturation (if any) in the sample.*
>
> *Solution: Again, we first convert the measured total mass fraction to a total concentration using Equation (2.32). This gives a value of 0.024 kg/m³ for C_T^{c-tet}. Because $\bar{C}_T^{c-tet}$ was previously calculated to be 2.95 kg/m³, $C_T^{c-tet} \ll \bar{C}_T^{c-tet}$, and no NAPL is present in this sample. The equilibrium gas and aqueous concentrations may be easily computed using Equation (2.54), with the saturated vapor concentration and aqueous solubility from the previous problem:*
>
> $$C_a^{c-tet} = (937)\frac{0.024}{2.95} = 7.6\,mg/l$$
>
> $$and\ \ C_w^{c-tet} = (755)\frac{0.024}{2.95} = 6.1\,mg/l$$

Mariner et al. [1997] and Mott [1995] describe computer programs that make similar calculations for the more complicated case of a multicomponent chemical system.

2.3.5   Multiphase Transport Mechanisms

Chemicals are transported in the gas, aqueous, and NAPL phases by advection, molecular diffusion, and mechanical dispersion. Advection is often the dominant transport process, and it is simply the transport of a chemical with the bulk movement of the phase. The chemical mass flux due to advection is the product of the chemical mass concentration and the Darcy velocity:

$$\mathbf{F}_\beta^i = C_\beta^i \mathbf{q}_\beta \tag{2.55}$$

Molecular diffusion of chemicals in the gas phase is an important transport mechanism in the vadose zone. The standard formulation, valid for "dilute" concentrations of non-condensing vapors, uses Fick's Law of diffusion with a correction for the porous medium influence:

$$\mathbf{J}_a^i = -\phi S_a \tau_a D_a^i \nabla C_a^i \tag{2.56}$$

Here, $\mathbf{J}_a^i$ is the gas phase diffusive mass flux, D_a^i is the chemical diffusion coefficient in free gas, and τ_a is called the tortuosity factor. The product $\phi S_a \tau_a$ is used to account for the reduced cross-sectional area available for diffusion in multiphase porous media, and for the longer, more tortuous diffusion paths in multiphase porous media. A similar formulation may be used to describe diffusion in the aqueous and NAPL phases:

$$\mathbf{J}_\beta^i = -\phi S_\beta \tau_\beta D_\beta^i \nabla C_\beta^i \tag{2.57}$$

The liquid molecular diffusion coefficients are about 10,000 times smaller than the gas phase diffusion coefficient. So, liquid diffusion processes are only important at smaller, local scales. Liquid diffusion plays an important role in the local kinetic interphase mass transfer of chemicals, where it often dominates the mass transfer process.

The gaseous diffusion of condensable vapors such as water vapor and some NAPL chemical vapors can be enhanced by local condensation and evaporation of the vapors across liquid films. This phenomenon is called "enhanced vapor diffusion," and it is important in some systems [Ho and Webb, 1998].

Mechanical dispersion is a diffusion-like spreading of chemical in a phase due to small-scale velocity variations. Dispersion theory is widely used in groundwater studies of dissolved matter, where the dispersive flux is used to account for chemical spreading due to velocity variations at length scales smaller than the measurement scale. The standard formulation for mechanical dispersion assumes a Fickian diffusion process, using an anisotropic, velocity-dependent dispersion coefficient. For water flow in the x-direction, the longitudinal (in the direction of flow), dispersive mass flux is calculated by

$$J_{w,l}^i = -D_{w,l} \frac{\partial C_w^i}{\partial x} \tag{2.58}$$

where the longitudinal dispersion coefficient is defined as

$$D_{w,l} = \alpha_l \frac{q_w}{\phi S_w} \tag{2.59}$$

The longitudinal dispersivity, α_l, depends on the scale of transport, as well as on the measurement scale of the velocity field. If the velocity field is determined at a sufficiently small scale, then the value of α_l needed to fit experimental data is relatively small. The dispersive spreading in the direction transverse to the flow is characterized by the transverse dispersive mass flux:

$$J_{w,t}^i = -D_{w,t} \frac{\partial C_w^i}{\partial y} \quad \text{or} \quad J_{w,t}^i = -D_{w,t} \frac{\partial C_w^i}{\partial z} \tag{2.60}$$

where the transverse dispersion coefficient is defined as

$$D_{w,t} = \alpha_t \frac{q_w}{\phi S_w} \tag{2.61}$$

and α_t is called the transverse dispersivity. The transverse dispersivity is typically five to twenty times smaller than the longitudinal dispersivity [Fetter, 1999]. The components of mechanical dispersion can be lumped into a single dispersion tensor, $\mathbf{D}$, such that the three-dimensional mass flux due to mechanical dispersion can be described as

$$\mathbf{J}^i_{\beta,m} = -\phi S_\beta \mathbf{D} \cdot \nabla C^i_\beta \tag{2.62}$$

The interested reader is referred to Bear [1979] for a complete description of the dispersion tensor. A similar mathematical formulation could be used to describe dispersion in the gas and NAPL phases. However, the full dispersion tensor for gas and NAPL phases is rarely, if ever, used in multiphase flow problems.

The transport mechanisms, advection, diffusion, and mechanical dispersion can be assembled into a mass balance equation for each component:

$$\frac{\partial(\phi S_\beta C^i_\beta)}{\partial t} = -\nabla \cdot (\mathbf{F}^i_\beta + \mathbf{J}^i_\beta + \mathbf{J}^i_{\beta,m}) \tag{2.63}$$

and substituting Equations (2.55), (2.57), and (2.62) into Equation (2.64) yields

$$\frac{\partial(\phi S_\beta C^i_\beta)}{\partial t} = -\nabla \cdot (C^i_\beta \mathbf{q}_\beta - \phi S_\beta \tau_\beta D^i_\beta \nabla C^i_\beta - \phi S_\beta \mathbf{D} \cdot \nabla C^i_\beta) \tag{2.64}$$

For a system containing all fluid phases, a system of three PDEs resembling Equation (2.64) must be solved for each chemical component i. The PDEs are coupled if phase partitioning is involved. If adsorption to the solid phase is significant, this phenomenon must be accounted for either with retardation coefficients such as that described in Equation (2.47) in the case of equilibrium adsorption, or with an additional PDE in the case of nonequilibrium conditions. Numerical solutions to Equation (2.64) are computationally intensive, due to the potentially large number of PDEs that must be solved and numerical restrictions such as those on time step sizes and grid block sizes.

3

Migration and Distribution

3.1 RESIDUAL AND TRAPPED SATURATIONS

Summary: Flowing NAPL will leave traces of residual phase due to immobilization. At the pore scale, residual NAPL exists as lenses (primarily in the vadose zone) or discontinuous blobs trapped by capillary forces (primarily in the saturated zone). At the field scale, geological heterogeneity is the controlling factor for entrapment. The residual NAPL is very difficult to recover and constitutes a long-term source of contamination, due to mass transfer to flowing water and air.

In the previous chapter, we presented the general mechanisms and a mathematical framework for analyzing NAPL flow in the subsurface. In many field cases, NAPL releases go often undetected until long after the source of the release is exhausted, the NAPL has spread to its maximum extent, and the dynamic flow phase of the NAPL contamination has ceased. The distribution of the NAPL is determined by the configuration of the release, the geological setting, and the physical-chemical properties of the NAPL, among other factors.

As discussed previously, NAPL will most commonly be the phase of intermediate wettability in the unsaturated zone. Under such conditions, all grain particles and contact points will be covered by water and NAPL will reside between water and air phases. Residual saturation is retained by capillary forces as films, as pendular rings at contact points between grains, and as droplets [Charbeneau, 2000]. An illustration of NAPL residing as a film is shown in Figure 3.1(a). Table 3.1 lists residual saturations in the unsaturated zone for a range of soil types.

In the saturated zone, NAPL immobilization occurs when the water imbibes back into the pore space previously occupied by the flowing NAPL and part of the NAPL is isolated from the NAPL body via bypassing and snap-off mechanisms [Wilson et al., 1990]. The residual NAPL is found as isolated droplets and ganglia. Figure 3.1(b) illustrates NAPL residing as a droplet in a pore space. Any pressure gradient that might exist in the NAPL body does not affect the droplets and ganglia, since the residual NAPL has been isolated

Soil and Groundwater Contamination: Nonaqueous Phase Liquids
Water Resources Monograph 17
Copyright 2005 by the American Geophysical Union
10.1029/17WM03

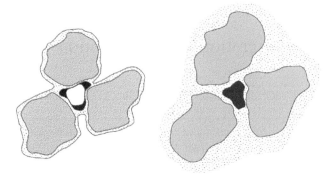

Figure 3.1. Illustration of different types of residual saturations: (a) residual NAPL saturation in the unsaturated zone (NAPL as the intermediate wetting fluid), (b) residual saturation in the saturated zone (NAPL as the non-wetting-fluid).

from the NAPL body. NAPL residual saturations in the saturated zone (see Table 3.1) tend to be higher than in the unsaturated zone, since the NAPL is trapped in the larger pores under liquid-saturated conditions. The wide range of residual saturations shown in Table 3.1 demonstrates that residual saturations are highly sensitive to soil properties. In general, the magnitude of residual saturations is considerably less dependent on fluid properties than on soil properties [Illangasekare, 1998].

TABLE 3.1.   Representative values for residual saturations for various soils in the unsaturated zone (NAPL as the intermediate wetting fluid) and the saturated zone (NAPL as the non-wetting fluid) [Parker et al., 1995].

Soil Type	ϕ	K_s (m/d)	Residual Saturation	
			Unsaturated Zone	Saturated Zone
Sand	0.43	712.8	0.03	0.26
Loamy sand	0.41	350.2	0.05	0.24
Sandy loam	0.41	106.1	0.05	0.23
Loam	0.43	25.0	0.07	0.19
Silty loam	0.45	10.8	0.07	0.17
Sandy clay loam	0.39	31.4	0.06	0.22
Clay loam	0.41	6.2	0.07	0.13
Silty clay loam	0.43	1.7	0.06	0.10
Sandy clay	0.38	2.9	0.07	0.10
Silty clay	0.36	0.5	0.04	0.05

3.2   THE ROLE OF GEOLOGICAL HETEROGENEITY

Geological heterogeneity has a dominant impact on the migration and entrapment patterns of NAPL. Contrasts in hydraulic conductivity at scales ranging from millimeters to meters can produce highly variable patterns of NAPL distributions. This situation is complicated by the fact that the impact of hydraulic conductivity contrasts can be different in the unsaturated and saturated zones. For example, NAPLs may be diverted around high-permeability strata in the unsaturated zone, but may be attracted to (and trapped in) high-permeability strata in the saturated zone. In this section, we demonstrate and develop explanations for the impacts of porous media heterogeneity on NAPLs.

The saturation levels indicated in Table 3.1 are representative for homogeneous geological units. However, as a rule, the subsurface is heterogeneous due to, for example, layering or lenses of looser or denser packing. There is abundant laboratory and field evidence that geological heterogeneity has a dominant impact on the migration and entrapment patterns of NAPL at scales ranging from millimeters to meters (see, e.g., Illangasekare [1998], Poulsen and Kueper [1992]).

Figure 3.2 shows some examples of the effects of geologic heterogeneities on NAPL distribution. On the one hand, low-permeability materials may inhibit downward migration and force NAPL to move laterally as demonstrated for LNAPLs and DNAPLs in unsaturated zone in Figure 3.2(a) and (c), respectively. Low permeability materials would act as barriers in the unsaturated zone only if these materials were fully saturated with water before the NAPLs were released. For example, clay lenses in the unsaturated zone would be expected to be fully saturated with water, except for especially dry environments. On the other hand, low-permeability materials that are only partially saturated with water can actually act

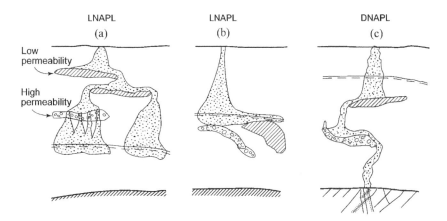

Figure 3.2. NAPL spreading and entrapment due to geological heterogeneity.

as sinks for NAPLs. The NAPL will "drawn" into the low-permeability material if, as is usually the case, the NAPL is nonwetting with respect to air and a fully-connected network of air within the pores of the low permeability material exists.

High-permeability units in the unsaturated zone may produce capillary barrier effects, which result in a temporary halt in the downward migration and lateral movement at the top of the coarser unit. After some time, the air-NAPL capillary pressure may be reduced to a sufficiently low value to allow the NAPL to enter the coarse layer. NAPL migration in the coarse layer will often take place as finger flow, as shown in Figure 3.2(a). Below the water table, however, a coarser sand lens may act as a trap for the NAPL, as shown in Figure 3.2(b) and Figure 3.2 (c). NAPL will easily enter the coarse lens, but will not enter the lower-permeability strata beneath until the NAPL-water capillary pressure accumulates to the point of exceeding the displacement pressure of the lower-permeability material.

The consequence of geological heterogeneity is that, at scales larger than the pore scale, the distribution of residual NAPL is heterogeneous and residual NAPL saturations cannot be estimated by simply integrating residual saturations for homogeneous materials (such as those indicated in Table 3.1) over different geological units. Examples shown in Figure 3.2 include potentially higher than expected saturations on the top of either low permeability units in a high permeability matrix (referred to as "pooling") or lower than expected saturations within high permeability units surrounded by a lower permeability matrix. A proper assessment of the extent and magnitude of a contamination amenable to remediation requires that the geological settings be carefully evaluated.

In Figure 3.3 through Figure 3.6, photographs from laboratory experiments conducted in tanks with transparent sides demonstrate the impacts of various types of geological heterogeneities on NAPL migration and distribution. In the first experiment, a 1-m high by 2-m wide by 5-cm deep "two-dimensional" tank was divided into two symmetric halves. The left half of the tank was packed with a coarse-grained sand to create a homogeneous packing configuration. The right half of the tank was packed with the same sand as on the left, except that a simple heterogeneity was created by embedding a block of fine sand within the coarse sand. After packing the tank, the soil was fully saturated by raising the water table to the top of the soil surface. The water table was dropped and the water was allowed to drain to create unsaturated conditions on both sections of the tank.

Equal volumes of a dyed NAPL (Soltrol) were released at the same rate on both halves of the tank. The photographs in Figure 3.3 show the propagation of the NAPL front and the final entrapment. The photographs in Figure 3.3 clearly show the impact of the fine sand block on the NAPL propagation and final entrapment. In this case, the fine material is only partially filled with water; the remainder of the pore space is filled with air. The NAPL is "drawn" into the finer material because it is non-wetting with respect to the air. Only air is displaced by the NAPL; the water remains at the irreducible saturation that existed before the NAPL was introduced. The NAPL residual saturation is higher in the finer material, so that a

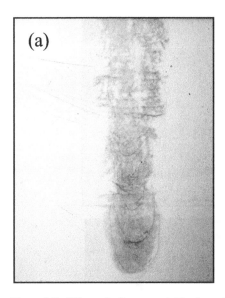 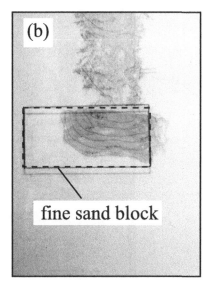

Figure 3.3. Effect of a fine material in the vadose zone on NAPL flow: (a) homogeneous, coarse sand packing and (b) fine sand block packed within coarse sand.

larger amount of NAPL is trapped in the finer material, relative to the amount of NAPL trapped in an equivalent volume of the coarse-grained sand.

In the second experiment, a two-dimensional tank was filled with a coarse-grained sand. A horizontal layer of a fine-grained sand was placed within the coarse sand in the tank below the water table. An amount of 2.0 kg of a dyed DNAPL (TCA or 1,1,1-trichloroethane) was released below the water table during a 40-minute period. Figure 3.4(a) shows the DNAPL migrating downwards due to gravitational forces. Figure 3.4(b) shows that, once the NAPL front reached the coarse/fine sand interface, the NAPL accumulated at the interface due to capillary barrier effects. The entry pressure of the finer sand is higher than that of the coarse sand. Most of the NAPL remained pooled at the interface but conditions can develop (based on the fluid properties and micro-scale heterogeneities) for the NAPL to finger through the finer soil. In Figure 3.4(c), it can be seen that the NAPL moved below the interface through fingering. Although all attempts were made to create homogenous packings of each material, the two homogeneous zones contained micro-scale heterogeneities. These micro-scale heterogeneities resulted in the staggered edges in the plume, as seen in Figure 3.4(b) and Figure 3.4(c).

In the third experiment, a two-dimensional tank was filled with a fine-grained sand. A horizontal layer of a coarse soil was placed within the fine sand in the tank below the water table. A mass of 3.9 kg of a test DNAPL (TCA) was released during a 68-minute period. Figure 3.5(a) shows the DNAPL migrating downwards due to gravitational forces. When the DNAPL front reached the

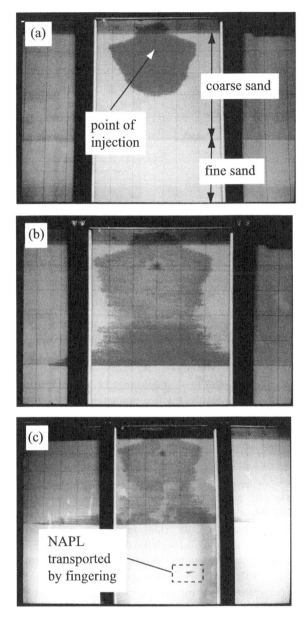

Figure 3.4. Capillary barrier effects at a coarse/fine sand interface.

fine/coarse sand interface, the DNAPL migration slowed until a finger(s) formed
(not visible through the tank wall) and drained the plume to the coarse/fine sand
interface, where it started to pool (Figure 3.5(b)). Again, micro-scale heterogeneities produced a slightly irregular edge to the DNAPL plume.

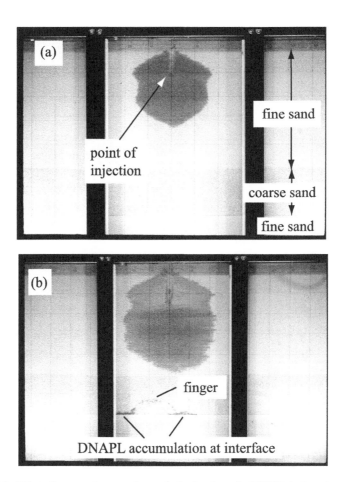

Figure 3.5. Effect of a coarse layer on the vertical migration of a DNAPL in the saturated zone.

In the fourth experiment, a two-dimensional tank was filled with a fine-grained sand. A sloping layer of a coarse soil was placed within the fine sand in the tank below the water table Figure 3.6. A test DNAPL was spilled below the water table. Figure 3.6 shows DNAPL migrating downwards due to gravitational and capillary forces. The DNAPL entered the coarse, inclined layer and migrated along continuous, micro-scale heterogeneities within the coarse, inclined layer. The DNAPL did not penetrate the bottom of the coarse, inclined layer because of the presence of the finer material at the bottom. Gravitational forces moved the DNAPL through the coarse, inclined layer until the DNAPL reached the end of the tank. At this point, enough DNAPL accumulated to produce pressures higher than the displacement pressure of the finer material, migrate vertically, and eventually form a pool at the bottom of the tank.

inclined layer
of coarse
material

Figure 3.6. DNAPL migration along an inclined capillary barrier.

3.3   INTERACTIONS WITH THE WATER TABLE

3.3.1   LNAPL Accumulation on Water Table

*Summary: For a LNAPL release in the unsaturated zone, the water table acts
as a hydraulic barrier for downward migration due to buoyancy forces. The
LNAPL will tend to accumulate on the top of the capillary fringe and spread
in lateral direction. If a sufficient amount of LNAPL accumulation takes place,
an LNAPL table may develop. Eventually, the displacement pressure for NAPL
entering into water saturated porous medium may be exceeded, allowing the
LNAPL to penetrate the saturated zone.*

A LNAPL release of sufficient size will penetrate the unsaturated zone and
reach the saturated zone. As the LNAPL moves towards the saturated zone, it will
encounter increasing water saturation and thus a reduced relative permeability.
Ultimately, the water table will tend to act as a hydraulic barrier for further downward migration. Initially, the LNAPL will pond on the barrier. But, with time, the
LNAPL may eventually overcome the buoyancy forces and dip beneath the initial

level of the water table. Concurrently, horizontal spreading takes place predominantly in the direction of the flowing groundwater. After the source is exhausted, the supply of LNAPL will no longer be sufficient to overcome the buoyancy forces and the LNAPL present in the saturated zone will undergo displacement by groundwater, leaving only the trapped fraction behind. As a result, the free LNAPL will be distributed over a limited vertical horizon above the water table, at least under idealized conditions.

Figure 3.7 shows the results of an experimental study of immiscible multiphase flow in a plexiglass laboratory groundwater flume packed with a homogeneous sand [Butts, 1996]. The water in the flume was manipulated to establish a sloping water table. A synthetic oil with a density less than water was injected at a constant rate at a point located in the unsaturated zone for a period of 2 hours and 20 minutes. The migration of the LNAPL plume was monitored by a video camera so that the front of the NAPL plume could be monitored with time. The results in Figure 3.7(a) show that before the LNAPL injection ceased, much of the LNAPL remained in the unsaturated zone and lateral migration occurred

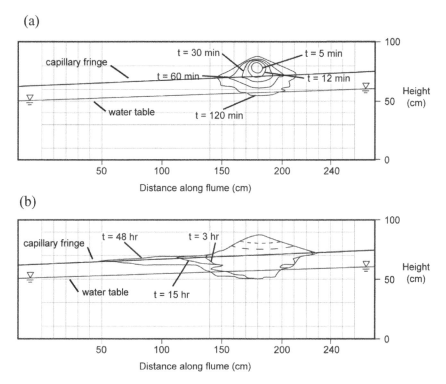

Figure 3.7. Distribution of the LNAPL plume after (a) 5, 12, 30, 60, 120 minutes and (b) 3, 15, 48 hours. The dashed lines indicate the LNAPL drainage front after 3 and 15 hours, [Butts, 1996].

primarily above the capillary fringe. After the LNAPL injection ceased a drainage front developed from the point of injection (dashed contour lines in Figure 3.7(b)). The majority of the LNAPL eventually migrated to a location above the capillary fringe and spread laterally over the capillary fringe in the direction of the slope of the water table. Figure 3.7(b) also shows that, at the location below the injection point, some water was displaced by NAPL from the capillary fringe and from below the water table. This phenomenon occurred even though the LNAPL is lighter than water since the height of the NAPL accumulating at this location was sufficient to overcome the capillary and buoyancy forces.

3.3.2 DNAPL Pooling on vs. Penetration of the Water Table

Summary: Because of higher densities compared to water, DNAPLs spilled at or near the ground surface, after reaching the saturation zone, will penetrate the water table and displace the water from saturated soil pores. The entry pressure of the soil controls this displacement process.

In this section, we discuss the mechanisms by which DNAPLs enter the saturated zone. To obtain a basic understanding of the mechanism of DNAPL entry into the saturated zone of an aquifer, let us first consider a simple case of a DNAPL spill in a homogenous formation. We assume that the DNAPL is non-wetting and water preferentially wets soil grains. In this discussion, we will consider a small point spill that corresponds to a leaking storage tank, although laboratory studies have shown that the manner in which DNAPLs move through the unsaturated zone depends on the size of the spill (e.g. Schwille [1988], Illangasekare et al. [1995a]). We also note that in most field situations, small-scale heterogeneities will cause the infiltrating DNAPL to form fingers and pools, rather than behave as a continuous body.

Figure 3.8 shows the saturation distribution when the DNAPL front is within the unsaturated zone, soon after the initial release. The capillary forces at DNAPL/air interfaces and the density and the viscosity of the DNAPL control the degree of lateral spreading. For example, a DNAPL with a low viscosity will penetrate the unsaturated soil very quickly. Depending on the source configuration and the depth to the water table, two scenarios can occur when the NAPL front reaches the water table. In the first scenario, we assume that we have a continuous DNAPL source and a shallow water table. In this scenario, shown in Figure 3.9, the DNAPL will accumulate and, consequently, the DNAPL head will build up above the water table. Within the capillary fringe, the water in soil pores is close to full saturation and is at pressures that are less than atmospheric. The DNAPL saturation is very low at the front when the front reaches top of the capillary fringe. This low DNAPL saturation corresponds to a very low relative permeability of the DNAPL phase.

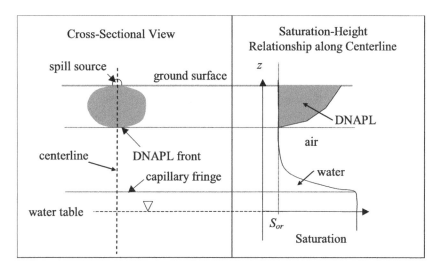

Figure 3.8. DNAPL migrating through the unsaturated zone: migration above the water table and capillary fringe.

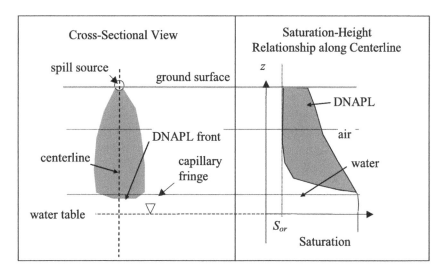

Figure 3.9. DNAPL migrating through the unsaturated zone: DNAPL reaches capillary fringe.

However, this situation does not last long, as the head in the accumulating DNAPL allows the DNAPL to displace the water in the capillary fringe, thus increasing the NAPL saturation. The capillary fringe collapses due to smaller capillary forces at the DNAPL/water interface within the capillary fringe, as compared to the capillary forces at the DNAPL/air interface that existed when the DNAPL was in the unsaturated zone. The driving DNAPL head in combination with the higher relative NAPL permeability pushes the DNAPL front through the capillary fringe. The water in the capillary fringe is displaced easily as the suction at the DNAPL/water interface contributes to the driving force at the front. With the penetration of the DNAPL front, the initial position of the water table (defined as the surface over which the gage water pressure is zero) moves downward. The net effect is that the DNAPL front moves into the saturated zone of the aquifer. In this scenario, we are assuming that the conditions at the DNAPL front are not unstable to produce fingering.

The second scenario involves a DNAPL source that is discontinuous in time such as in the cases of a slug application or a deep water table. This scenario, shown in Figure 3.10, produces immobile and mobile NAPL regions within the saturation zone. Within the immobile region, the DNAPL is found in the form of discrete blobs or ganglia (residual saturation). The DNAPL in the mobile region is continuous and moving. For the non-wetting NAPL to enter water-saturated pores below the water table, it has to displace the water that is at a pressure higher than atmospheric. The pressure that is needed at the DNAPL front to displace the water is the displacement pressure of the aquifer material. Because the DNAPL density is higher than that of water, when a sufficient depth of DNAPL accumulates, the capillary pressure across the DNAPL/water interface exceeds the displacement pressure of the soil and DNAPL enters the saturated zone of the aquifer through gravity-driven migration. However, the front may not remain sharp, if conditions at the front are instable, resulting in finger initiation (see Section 3.5.1).

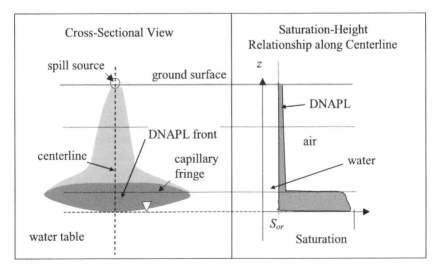

Figure 3.10. DNAPL migrating through the unsaturated zone: DNAPL enters saturated zone.

3.3.3 Smearing of LNAPL Due to Water Table Fluctuations

Summary: Water table fluctuations may have a significant impact on the spatial distribution of a LNAPL plume residing on the top of a water table. The fluctuations may lead to a larger fraction of the plume being trapped as residual saturation so that a smaller fraction can be recovered by hydraulic means.

When the water table falls, LNAPL overlying the water table will follow the movement of the water table. As the LNAPL drains from the pore space to follow the water table, a trace of residual saturation will be left behind, as shown in Figure 3.11. As a result, a larger fraction of the spill is smeared over a horizon in the unsaturated zone and amount of mobile LNAPL will decrease. Conversely, if the water table rises, the overlying LNAPL will rise with the water table and residual LNAPL will be left in the saturated zone. We should note that the residual LNAPL above and below the capillary fringe has been trapped by two different mechanisms. LNAPL is trapped in the vadose zone as thin films; in the saturated zone, it is trapped as disconnected ganglia.

Figure 3.12 shows a map of a gasoline service station in Michigan where a 1,500-liter tank containing gasoline ruptured catastrophically. Beginning approximately one year after the tank ruptured, 18 months of water table and LNAPL table elevation data were collected. These data are shown in Figure 3.13 and Table 3.2. No LNAPL recovery efforts were undertaken during this period. The elevations indicated in Figure 3.13 are average elevations from four monitoring wells located near the aerial center of the LNAPL plume. The LNAPL thicknesses given in Figure 3.13 are estimated by subtracting water elevations from LNAPL elevations in wells. The results in Figure 3.13 show that the LNAPL elevations follow the seasonal water table fluctuations closely. Figure 3.13 also shows that the NAPL thickness decreases with time, indicating that the amount of mobile LNAPL in the vicinity of the monitoring wells is decreasing. This observation is confirmed by the soil boring data in Table 3.2, where we see that the vertical extent of the LNAPL has decreased over time. The decrease in mobile LNAPL is most likely caused by NAPL smearing by the rise and fall of the water table, which has produced a larger fraction of residual LNAPL. Some of the decrease in mobile LNAPL may also be due to lateral redistribution of the LNAPL. A quantitative analysis of the amount of mobile LNAPL, based on LNAPL well thickness, is given in Section 4.1.1.

Figure 3.14 shows how the loss of NAPL to residual can be explained by following the travel along the drainage and imbibition paths of the capillary pressure curves. Figure 3.14 (a) illustrates a situation corresponding to hydrostatic conditions. Some of the NAPL is retained in the upper part of the unsaturated zone as residual NAPL. The remaining NAPL is present in the formation as mobile NAPL and corresponds to the height of NAPL in an observation well. In accordance with the common wettability assumptions, the water saturation profile is determined by the water-NAPL capillary pressure curve and the total liquid saturation

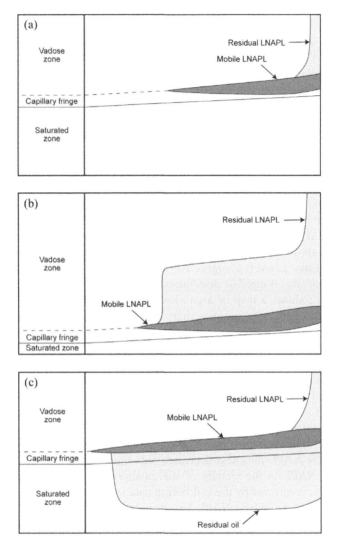

Figure 3.11. Effect of falling and rising water table on the distribution of mobile and residual phases of an LNAPL. Fetter, C.W., Contaminant Hydrogeology, 2nd Edition, 1999. Reprinted by permission of Pearson Education, Inc., Upper Saddle River, NJ.

(and thus the NAPL saturation) profile is determined by the NAPL-air capillary pressure curve. We assume that the fluid distributions can be represented with drainage curves.

If the water table falls, the mobile NAPL will follow the movement of the water table and the trace of residual NAPL will be extended. A new hydrostatic

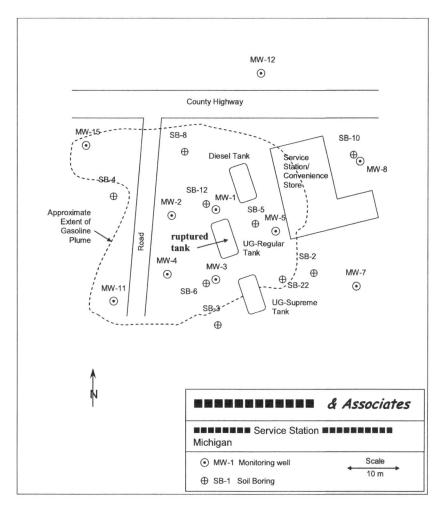

Figure 3.12. Site map for service station with LNAPL release.

situation and a new set of saturation profiles will develop. The two drainage capillary pressure curves are similar to those shown in Figure 3.14 (a) but are now
shifted downwards (Figure 3.14 (b)). The amount of free NAPL is reduced by the
additional loss to residual. The thickness of NAPL in the monitoring well is
reduced by the same amount.

For a subsequent rise in the water table, imbibition of water into the liquid-
saturated zone (no air present) and imbibition of NAPL into the liquid-unsatu-
rated zone (air present) will occur. A new hydrostatic situation will eventually
arise where the saturation profiles are determined by the imbibition paths of the
capillary pressure curves (Figure 3.14 (c)). As a result, NAPL will be trapped in

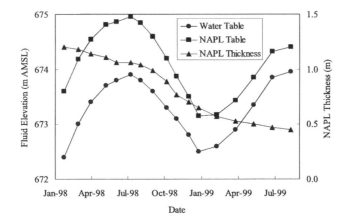

Figure 3.13. Water and LNAPL elevation observations and estimated NAPL thickness from monitoring wells at a gasoline station. Observations are averages over four monitoring wells near the center of the LNAPL plume (MW-1, MW-2, MW-3, MW-5; see Figure 3.12 for location).

the zone previously containing only water and air will be trapped in the zone previously containing only water and NAPL. It is also important to note that, as the saturation profiles are shifted upwards, part of the residual NAPL is now converted to a continuous fluid that can be mobilized.

If the water table fluctuates within a zone where NAPL smearing has occurred, the scenario becomes more complicated because of the differences between residual NAPL saturations above and below the water table. For a falling water table, the unsaturated zone residual NAPL increases, but, at the same time part of the saturated zone residual is recovered. The amount of continuous LNAPL increases because the saturated zone residual is generally larger than the unsaturated zone residual. As a result, greater thicknesses of LNAPL may be observed in monitoring wells when the water table falls.

TABLE 3.2.   Vertical extent of NAPL as determined from soil borings (SB), see Figure 3.12 for location.

Date	Vertical extent of NAPL as Determined from Soil Boring (m)		
	SB-12	SB-5	SB-6
Feb-98	1.24	1.08	0.86
Jan-99	1.55	1.77	1.34
Jul-99	2.07	1.93	1.55

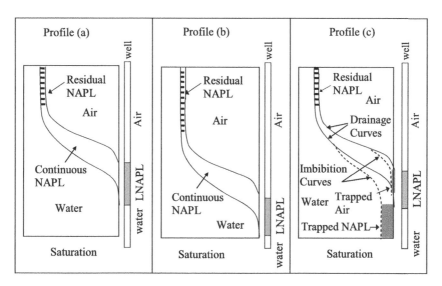

Figure 3.14. Change in saturation profiles and fluid levels in monitoring wells due to a falling and rising water table.

3.4 CAPILLARY BARRIERS AND CHANNELS

3.4.1 NAPL Pooling on vs. Penetration of Low Permeable Material

Summary: All underground porous formations in their natural state will exhibit variability of properties in space. As DNAPLs penetrate the saturated zone, they encounter these heterogeneities. The flow behavior controlled by these heterogeneities will ultimately determine the distribution and entrapment of DNAPLs in the subsurface. The conditions under which a DNAPL will enter a finer soil at the interface of a coarse/fine layer are determined by a critical entry pressure. Until this critical pressure is reached, the DNAPL will pool at the interface. Expressions to determine the DNAPL saturation distribution within a pool are developed.

In natural soil formations, the movement of contaminants is complicated by soil heterogeneities such as fractures, macropores, and layering. Experimental studies by Schwille [1988], Kueper and Frind [1991], and Illangasekare et al., [1995a, 1995b], have shown that soil layering can cause lateral spreading, preferential flow, and pooling of the organic liquid. In this section, we investigate the mechanism of DNAPL pooling at the interfaces of formations with contrasting soil properties.

After a DNAPL enters the saturated zone, it may encounter interfaces between two soil layers. Pooling at an interface is a result of two mechanisms that may work in conjunction to inhibit the movement of the NAPL from a coarse to a finer soil layer. First, for the NAPL to enter from the coarser soil to the finer, the NAPL must develop a sufficient head to overcome the larger entry pressure for the finer soil. This mechanism is known as the capillary barrier mechanism and is most often referred to as the primary mechanism responsible for pooling. The second mechanism that inhibits the movement of the NAPL is controlled by the permeability of the NAPL in the finer soil. Even with a high driving head, a lower permeability results in a high resistance to flow into and pooling on top of the finer material.

We note that, although both the capillary barrier mechanism and the permeability mechanism can be responsible for pooling, the physics behind the two mechanisms are different. The capillary barrier mechanism is an "on-off" mechanism, that is, if sufficient head for penetration is not developed, the NAPL will not penetrate the finer material, and vice versa. The permeability mechanism controls the rate of flow, such that at very low permeabilities the NAPL may actually be flowing into the finer material, but so slowly that even over long time periods, the NAPL appears to be immobilized on the top of the finer material.

As the permeability mechanism is well understood, in the following, we further explain the capillary barrier mechanism and develop an expression for the minimum capillary head that is needed at the interface for the NAPL to enter a finer soil. In the case of a non-wetting DNAPL displacing the wetting water phase, the capillary head must exceed a critical entry head for the fluid to enter from a coarse soil to a finer soil. To demonstrate and analyze this effect, a series of experiments were conducted (Fairbanks, [1993], Illangasekare [1998]) where a NAPL was pumped into a water saturated heterogeneous soil column. Changes in water saturation due to NAPL displacement were observed and the drop in pressure across the NAPL-water interface (capillary pressure) was measured. The testing apparatus shown in Figure 3.15 consisted of a flexible wall permeameter and a flow pump. The system was also equipped with pressure cylinders for controlling fluid pressures, a differential pressure transducer for measuring capillary pressures, a ceramic high air-entry porous bottom plate and a data acquisition system. A precision, low-flow pump was used to withdraw water from the bottom of the sample through the porous plate. A reservoir containing a LNAPL was connected to the top cap of the cylinder.

Figure 3.16 shows a plot of capillary pressure versus the measured water saturation for case of a heterogeneous system consisting of a layer of coarse sand placed over a layer of finer sand (mean grain sizes d_{50} of 0.880 mm and 0.185 mm, respectively). The measured water saturation is the fraction of the total pore space of the heterogeneous soil sample occupied by water. The results in Figure 3.16 show that, as the water saturation decreased with the displacement of water by the NAPL, the capillary pressure increased. The capillary pressure in this case is the pressure drop across the sample measured by the differential transducer.

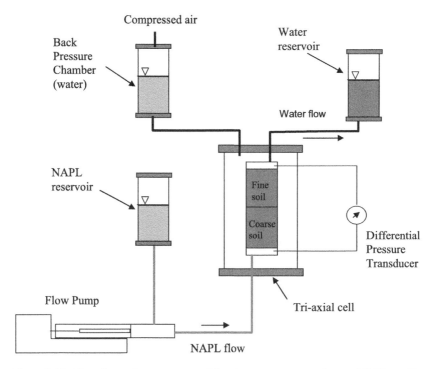

Figure 3.15. Experimental apparatus used for the measurement of water/NAPL capillary pressure-saturation curve for a layered soil sample.

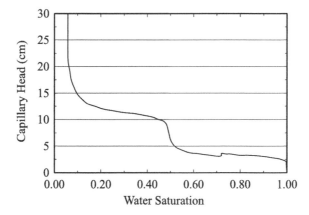

Figure 3.16. Representative curve of capillary pressure variation in a layered system.

As progressively smaller pores are de-saturated, the capillary pressure curve followed the water retention curve of the coarse sand. When the NAPL front reached the interface between the two soil layers, the capillary pressure increased abruptly, representing the higher pressure needed for the NAPL to enter the finer soil. The magnitude of the transition in capillary pressure depends on the contrast in the properties of the two soils.

To determine under what conditions a DNAPL will enter a soil at a coarse/fine soil interface, first let us derive a simple equation that could be used to estimate the capillary pressure distribution in a two-phase system (see idealized situation shown in Figure 3.17). For each fluid phase under hydrostatic equilibrium, the force balance equation can be written as,

$$-\frac{dp}{dz} - \rho g = 0 \tag{3.1}$$

where g is the gravitational acceleration, p is the pressure, ρ is the fluid density, and we take z as positive upward. Integrating Equation (3.1) and defining the pressure at the datum to be p_0, the pressure at any elevation can be expressed in the form,

$$p - p_0 = -\rho g z \tag{3.2}$$

Applying Equation (3.2) to the water and NAPL phases separately gives,

$$p_w - p_{w0} = -\rho_w g z \tag{3.3}$$

and

$$p_{nw} - p_{nw0} = -\rho_{nw} g z \tag{3.4}$$

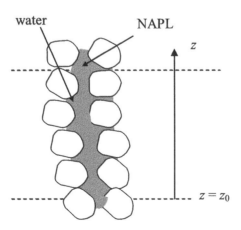

Figure 3.17. Static two-phase system.

By definition, the capillary pressure at NAPL/water interface under hydrostatic conditions is

$$p_c = p_{nw} - p_w \tag{3.5}$$

Subtracting Equation (3.3) from (3.4) and using Equation (3.5), an expression for the capillary pressure at any elevation in the two phase static system is obtained as

$$p_c(z) = (\rho_w - \rho_{nw})gz + p_{c0} \tag{3.6}$$

where p_{c0} is the capillary pressure at the datum.

Continuing to assume hydrostatic conditions, we can use Equation (3.6) to investigate the condition under which a static pool at a coarse/fine interface will promote entry of DNAPL into the finer material [Kueper and McWhorter, 1991]. Consider the case of a static DNAPL pool of thickness D at the interface of two soils as shown in Figure 3.18. In this case we will assume that the pool is in recession. Under these conditions, it could be assumed that the capillary pressure at the top of the pool is close to zero. If the depth of the DNAPL pool is D, then at $z = D$ the capillary pressure is:

$$p_c(D) = 0 \tag{3.7}$$

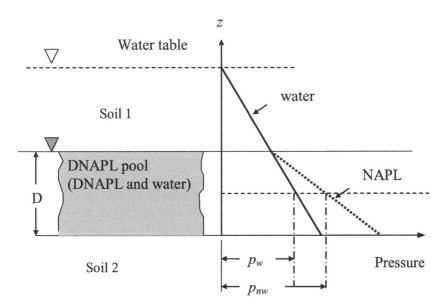

Figure 3.18. Pressure distribution within a DNAPL pool at two soil interfaces.

Substituting Equation (3.7) into Equation (3.6) and solving for p_{c0}, the capillary pressure at the interface between the two soils is:

$$p_{c0} = (\rho_{nw} - \rho_w)gD \tag{3.8}$$

As the DNAPL has already entered the soil 1, we have

$$p_{c0} > p_d^{(1)} \tag{3.9}$$

where $p_d^{(1)}$ is the displacement pressure of DNAPL in soil 1.

If the second soil is coarser than soil 1, then

$$p_d^{(1)} > p_d^{(2)} \tag{3.10}$$

where $p_d^{(2)}$ is the displacement pressure of soil 2. Thus, from Equation (3.9), we have

$$p_{c0} > p_d^{(2)} \tag{3.11}$$

Equation (3.11) implies that DNAPL will enter the coarser layer for any DNAPL depth greater than zero such that there will be no DNAPL pooling at the interface.

Now let us consider the case where the second soil is finer than the soil in the first layer. In this case

$$p_d^{(2)} > p_d^{(1)} \tag{3.12}$$

and for the DNAPL to enter soil 2, the capillary pressure at the interface must reach the value $p_d^{(2)}$. From Equation (3.8), it can be shown that for the DNAPL to enter soil 2, the pool depth should be at least

$$D_{min} = \frac{p_d^{(2)}}{(\rho_{nw} - \rho_w)g} \tag{3.13}$$

From Equation (3.13), it can be seen that (a) a denser DNAPL will enter the second soil layer more easily than a less dense DNAPL and (b) a DNAPL will pool readily at the interface when the second soil has a high displacement pressure typical of fine grained soils or clays.

Example: Consider the example of TCE with a density of 1.47 g/cm³ encountering a fine sand with a displacement pressure head of 0.250 m. Also consider a coal tar with a density of 1.05 g/cm³ that pools at the same soil interface.

Solution: Applying Equation (3.13) for the case of TCE, the TCE depth necessary to enter the finer layer is

$$D_{min} = \frac{0.250 \text{ m}}{(1,470 \text{ kg/m}^3 - 1,000 \text{ kg/m}^3)(9.80 \text{ m/s}^2)} = 0.532 \text{ m}$$

For the coal tar we have

$$D_{min} = \frac{0.250 \text{ m}}{(1,050 \text{ kg/m}^3 - 1,000 \text{ kg/m}^3)(9.80 \text{ m/s}^2)} = 5.0 \text{ m}$$

It should be noted that the saturation distribution within the DNAPL pool over the depth D is not constant. To determine the saturation distribution within the pool, first we will combine Equations (3.6) and (3.8) to obtain an expression for the capillary pressure distribution as

$$p_c(z) = (\rho_{nw} - \rho_w)(D - z)g \qquad (3.14)$$

Substituting for p_c from the Brooks-Corey model (Equation (2.6)) in Equation (3.14), we obtain [McWhorter and Kueper, 1996]

$$z = D - \frac{p_d}{(\rho_{nw} - \rho_w)g}\left[\frac{S_w - S_{wir}}{1 - S_{wir}}\right]^{-1/\lambda} \qquad (3.15)$$

Note that the above expression is valid for the zone within the DNAPL pool where the capillary pressure is higher than the displacement pressure p_d. A more general expression for the whole pool depth can be derived using the van Genuchten model, where it is assumed that the capillary pressure distribution is continuous (no transition displacement pressure). By substituting the expression for capillary pressure from van Genuchten model (Equation (2.8)) in Equation (3.14), McWhorter and Kueper [1995] derived

$$z = D - \frac{p_o}{(\rho_{nw} - \rho_w)g}\left[\left(\frac{S_w - S_{wir}}{1 - S_{wir}}\right) - 1\right]^{1-m} \qquad (3.16)$$

where p_o is a characteristic pressure and m is a fitting parameter from the van Genuchten model.

Equations (3.15) and (3.16) show that saturations within the pool are sensitive to the displacement pressure and the density difference between the DNAPL and water. McWhorter and Kueper [1995] showed that the ratio

$$\Gamma = \frac{D(\rho_{nw} - \rho_w)g}{p_0} \qquad (3.17)$$

controls the mode of saturation distribution within the pool, where significant saturation can exist in a pool when Γ is greater than unity.

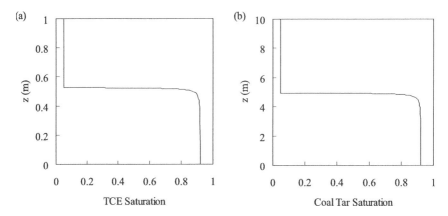

Figure 3.19. Saturation distribution in a (a) TCE pool and (b) coal tar pool above a #70/#125 sand interface in a layered system.

Example: Determine the saturation distribution within a hypothetical TCE ($\rho_o = 1.47$ g/cm^3) pool and a coal tar ($\rho_o = 1.05$ g/cm^3) pool above the interface between #70 and #125 sands. The depth of the TCE pool is $D = 0.531$ m and the coal tar pool depth is $D = 4.99$ m. Each of these depths is slightly above the displacement pressure ($p_d = 0.532$ m of water) for the #125 sand (see previous example). Other parameters for the #125 sand required to use the Brooks-Corey model to calculate the saturation distribution are $\lambda = 1.5$ and $S_{wir} = 0.08$.

Solution: Figure 3.19 (a) and (b) show the saturation distributions for the TCE and coal tar pools, respectively, calculated from Equation (3.15).

Illangasekare et al. 1995, conducted a set of experiments in intermediate-scale laboratory tanks to investigate the behavior of DNAPLs at the interfaces of layered soil formations. In these experiments, TCA (1,1,1-trichloroethane) with density 1,349 kg/m^3 was used as the test DNAPL. In the first experiment, a fine sand, #70, was used to create a layer within a #30 coarse formation. Saturation profiles along the vertical centerline of the flume were recorded using a dual gamma system at various time periods. Sample saturation profiles at 6.3 to 41.8 hours after the spill are shown in Figure 3.20.

Due to the driving head at the point of injection, the TCA initially spread in all directions. Because the density difference between water and the DNAPL is significant, the movement of the organic plume was mostly vertically downward through the #30 sand. Fifteen hours after the spill, the fine layer acted as a barrier and the fluid started accumulating at the coarse/fine sand interface. Substituting for the displacement pressure of #70 sand ($p_d^{(2)} = 1325$ N/m^2) and density of TCA ($\rho_o = 1,349$ kg/m^3) in Equation (3.13), the minimum depth, D_{min}, of TCA needed

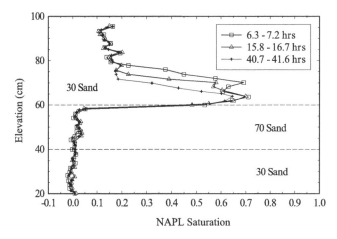

Figure 3.20. Saturation profiles of fine layer TCA experiments 6.3 to 41.6 hours after spill.

for displacement at the #30/#70 interface is 39 cm. As the pool depth did not reach this minimum value, the TCA did not enter the finer #70 sand and the TCA spread laterally at the interface of the two soils.

After the first three hours of injection, spreading of the DNAPL continued, but at a much slower rate. The saturation data of Figure 3.20 clearly shows that TCA continued to drain from the coarse layer. Since the DNAPL had practically ceased lateral spreading, but continued to migrate from the coarse layer, it must have moved through the fine layer as single or multiple fingers. Visual observation of the plume showed a finger developing in the #30 sand layer, beneath the #70 sand layer. The physics of NAPL finger formation is discussed in Section 3.5.1.

A second experiment was conducted where the fine #70 sand layer was replaced with a coarse #16 sand layer. Approximately one hour and 50 minutes into the spill, TCA fingers were seen developing in the coarse layer. The instability is believed to have been triggered by removal of the applied head when injection was discontinued. After fingering through the #16 sand layer, the TCA was initially unable to achieve the entry pressure of the #30 sand and began to pond on top of the #30 sand. Figure 3.21 presents the saturation profile after the TCA began to pond and spread at the base of the coarse layer. The TCA is shown moving from the #30 sand in upper portions of the flume and ponding at the base of the #16 layer.

Ponding led to the development of pressure gradients in the DNAPL phase and subsequent stable flow in the form of spreading along the base of the coarse layer. Ponding also raised the pressure of the TCA to the entry pressure of the #30 sand. This initiated unstable flow in the #30 sand and the gravel pack below. Fingers that developed in this experiment were much smaller than the thickness of the flume or the resolution of the dual-gamma system. Consequently, TCA saturation

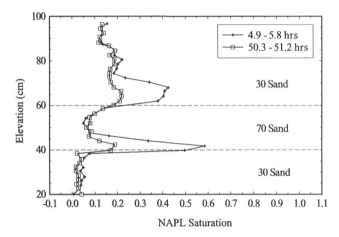

Figure 3.21. Saturation profiles of coarse layer TCA experiments 4.9 to 51.2 hours after spill.

data in the unstable flow areas does not represent the saturation of the fingers themselves, but does represent the macroscopic saturation. TCA saturations above 5% occurred at several locations where visual inspection of the flume indicated TCA was not present.

3.4.2   Migration of DNAPLs along Sloping Confining Layers

Summary: When a DNAPL encounters an interface between two soils, pooling can occur depending on contrasts in properties of soils in the two layers. However, if the barrier surface is not horizontal, gravity will move the DNAPL along the interface. This situation can occur in sloping confining layers and on less permeable bedrock formations. A simple capillary tube model is presented to understand the forces controlling the movement of a DNAPL over an inclined surface. This model shows how gravity and capillarity affect the rate of movement. The model is used to describe how a DNAPL slug moves over a confining layer.

Figure 3.22 (a) shows a DNAPL release that has migrated to a confining layer that is acting as a capillary barrier. The mobile portion of the release is migrates along a confining layer. In Figure 3.22(b) we picture only the mobile portion of the release, where it is assumed that the DNAPL source is inactive and a finite volume of the DNAPL is migrating along the confining layer. Two regions can be identified in the DNAPL/water interface. As the DNAPL migrates along the front portion of the boundary, the non-wetting fluid displaces the wetting water phase.

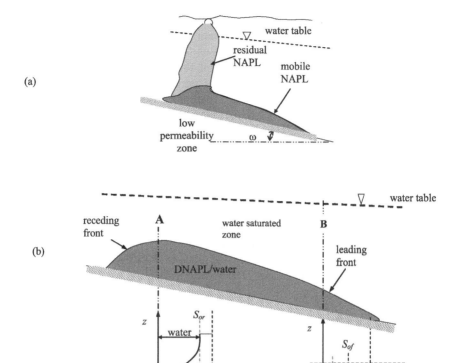

Figure 3.22. DNAPL migrating along a sloping surface.

In the tail of the DNAPL plume, the non-wetting phase recedes and the wetting phase occupies the available pore space. The displacing water phase cannot completely remove the non-wetting phase. This results in immobile DNAPL in the form of blobs and ganglia left behind the receding front.

In vertical section A, the front is receding. As shown in Figure 3.22, at this location the DNAPL saturation varies from saturation $(1 - S_{wir})$ at the confining layer surface to S_{or} at the top, where S_{wir} is the irreducible water saturation and S_{or} is the residual NAPL saturation. Since the saturation of DNAPL increases with depth, the relative permeability of the DNAPL also increases. That is, given the same gradient, the DNAPL will move faster in the regions close to the confining layer. Because of the low capillary pressures at the upper boundary (close to zero), NAPL migration will be primary controlled by gravity. The slow movement at the top boundary results in the NAPL becoming immobilized, leaving a residual zone of DNAPL entrapment.

At section B, at the leading edge, DNAPL saturations are relatively uniform along the vertical section. If the same gradient is similar along the depth of the front, the NAPL velocity will be more or less uniform along with depth. However, as the DNAPL becomes immobilized in the tail, the saturation at the leading front decreases, and correspondingly, the relative permeability of the non-wetting phase is reduced. This reduction in relative permeability results in a decrease in the front velocity as the slug is "stretched" along the confining later. Eventually, a final state will be reached where all the DNAPL will be at immobile saturation.

To obtain a conceptual understanding of how capillary and gravity forces control the movement of a DNAPL in a confining layer, we can examine the case of an inclined capillary tube along which a DNAPL front is propagating by displacing water (Figure 3.23). A DNAPL pool of depth h_0 is placed at the entrance to the tube. The top of the pool is at a depth h_w below the water table. The tube outlet is subjected to a water head of $h_w + \Delta h_w$. L is the length of the tube. Let l_f be the location of the DNAPL front at any time t.

The flow of a fluid under an external head gradient in a tube of radius r can be described using Poiseuille's equation. The average velocity, u, can be expressed as

$$u = -\frac{r^2}{8\mu}\frac{d}{dx}(p + \rho g z) \tag{3.18}$$

where p is the fluid pressure, ρ is the fluid density, μ is the fluid viscosity and z is the elevation head. Applying Equation (3.18) to the DNAPL phase flowing in the tube and recognizing that the DNAPL/water front velocity is the same as the fluid velocity ($u = dl_f/dt$), we can derive an expression for the rate of front propagation as

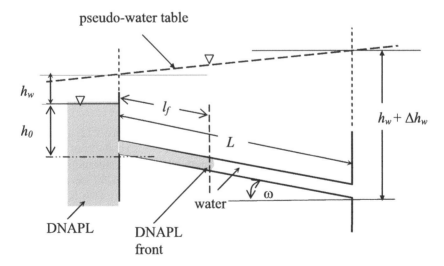

Figure 3.23. DNAPL migration in an inclined capillary tube.

$$\frac{dl_f}{dt} = -\frac{r^2}{8\mu_o}\frac{\left[p_o^f + \rho_o g(L - l_f)\sin\omega - (h_w\rho_w g + h_o\rho_o g + \rho_o g L\sin\omega)\right]}{l_f} \quad (3.19)$$

where μ_o is the viscosity of the DNAPL, ρ_o is the density of the DNAPL and p_0^f is the DNAPL pressure at the front. Similarly applying Equation (3.18) for water, we obtain

$$\frac{dl_f}{dt} = -\frac{r^2}{8\mu_w}\frac{\left\{(h_w + \Delta h_w)\rho_w g - \left[p_w^f + \rho_w g(L - l_f)\sin\omega\right]\right\}}{(L - l_f)} \quad (3.20)$$

By definition, the capillary pressure at the DNAPL/water interface is

$$p_c^f = p_o^f - p_w^f \quad (3.21)$$

Combining Equations (3.19), (3.20) and (3.21), we obtain

$$\frac{dl_f}{dt} = \frac{r^2}{8\left[\mu_o l_f + \mu_w(L - l_f)\right]}\Big\{h_o\rho_o g - \Delta h_w \rho_w g - p_c^f$$
$$+ \left[(\rho_o - \rho_w)l_f + \rho_w L\right]g\sin\omega\Big\} \quad (3.22)$$

Equation (3.22) can be simplified by introducing a water table slope, i_w, to define the water head drop, Δh_w, as in

$$\Delta h_w = i_w L\cos\omega + h_o + L\sin\omega \quad (3.23)$$

Substituting for Δh_w in Equation (3.22) and rearranging terms,

$$\frac{dl_f}{dt} = \frac{r^2}{8\left[\mu_o l_f + \mu_w(L - l_f)\right]}\left[(\rho_o - \rho_w)(h_o + l_f\sin\omega)g - p_c^f - i_w\rho_w g L\cos\omega\right] \quad (3.24)$$

The terms within brackets in the numerator on right hand side of Equation 3.24 identify the different driving forces that contribute to the propagation of the DNAPL front. The first term, $(\rho_o - \rho_w)(h_o + l_f \sin\omega)g$, represents the gravitational driving force resulting from the density difference and the length of NAPL that has entered the capillary tube. The second term, p_c^f, is a resisting force due to the capillary pressure. The third term is associated with the groundwater slope represents driving forces created by the water table. This term represents a resisting force if the DNAPL movement is in direction opposite to groundwater flow (the situation described in Figure 3.23), but is a driving force if the DNAPL movement is in the direction of groundwater flow. The numerator in the leading term, r^2, in Equation (2.32) indicates that the rate of propagation is proportional to the square of the pore size. The denominator in the leading term represents the viscous resistance to flow, such that the rate of propagation is inversely proportional to the viscosity of both fluids.

Example: A 0.5 m long tube with a radius of 0.002 m (4 mm diameter) is filled with an immiscible fluid with a density 1,587 kg/m³ and dynamic viscosity 9.77 × 10⁻⁴ N·s/m² and water in the configuration shown in Figure 3.23. The density and viscosity of water are 998.2 kg/m³ and 1.01 × 10⁻³ N·s/m², respectively. The surface tension between the DNAPL and water interface is 0.5 N/m. Determine the DNAPL/water interface velocity at the midpoint of the tube for the following tube angle and water table slopes (a) tube slope: ω = 15° and water table slope = + 0.01 (downward); (b) tube slope: ω = 15° and water table slope = −0.01 (upward); and (c) tube slope: ω = 0° and water table slope = 0.

Solution: (a) The front velocity when the DNAPL/water interface reaches the midpoint of the tube, as calculated from Equation (3.24), is 0.40 m/s. In this case, the gravity component of the driving head is 952 N/m² and the capillary component is 500 N/m². (b) The front velocity increases to 0.50 m/s. (c) The front velocity decreases to 0.08 m/s. The gravitational driving head is 578 N/m² and the capillary drive is 500 N/m². This numerical example demonstrates that, conceptually, the rate of propagation of a DNAPL front along sloping is controlled by the relative contribution of the three driving forces created by gravity, capillarity and the water table slope.

3.4.3   Occurrence of NAPLs in High Permeability Regions

Summary: NAPLs, after entering the saturated zone, can migrate preferentially through zones of high permeability or low displacement pressure. High permeability zones are found in natural aquifer formations or a result of back filling by coarser soils used in utility corridors. Within these high permeability zones, the NAPLs can migrate or remain trapped at saturations that are larger than the residual values. The mechanisms that control the occurrence of LNAPLs and DNAPLs in high permeability regions in aquifers are discussed and demonstrated with experimental data.

Coarser materials can occur in subsurface formations due to geologic stratification or due to human rearrangement of natural materials. For example, underground utility corridors containing pipes or cables are normally filled with coarser back fill materials. Because the entry pressures are low and permeabilities are high, NAPLs enter these layers easily and migrate preferentially. These preferential channels can transmit significant volumes of NAPLs. A potential scenario of a LNAPL moving preferentially through a coarse material underlying a finer material is shown in Figure 3.24, where tank containing a LNAPL is leaking into an unconfined aquifer. The hypothetical capillary pressure-saturation functions for the two materials are given in Figure 3.25. Note that soil 1 has a higher displacement pressure than soil 2. In this scenario, we assume that there is

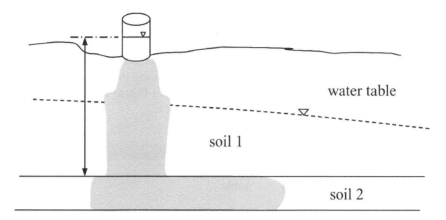

Figure 3.24. Preferential flow of LNAPL into a high permeability layer.

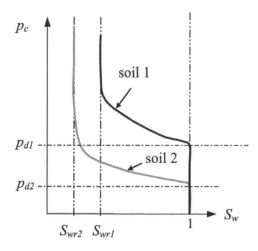

Figure 3.25. Retention functions of the two soils in the heterogeneous formation.

sufficient head in the tank for the LNAPL to penetrate the capillary fringe after migrating through the unsaturated zone. After entering the saturated zone, the LNAPL encounters the interface between soil 1 and soil 2 in the stratified system. Because the displacement pressure of the coarse soil 2 in the layer is smaller than that of soil 1, the LNAPL will displace the water and enter the coarser sand layer. The water table gradient as shown will cause the LNAPL to migrate from left to right.

Illangasekare et al. [1995a] demonstrated the preferential migration of a LNAPL in permeable materials in a series of experiments conducted in a large soil tank. In one of the experiments, a heterogeneous packing configuration was created in a 9.8 m-long, two-dimensional tank by inserting a layer of #16 coarse sand into a finer formation of #30 sand (see Figure 3.26). The water table was located above the coarse layer. The test LNAPL, Soltrol 220, was injected into the coarse layer. Transient saturation profiles recorded during the propagation of the LNAPL through the coarser sand layer are shown in Figure 3.27. The LNAPL

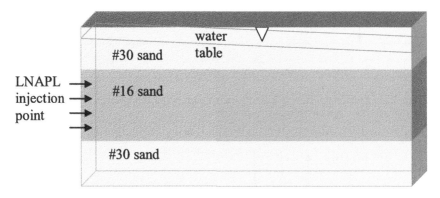

Figure 3.26. Schematic of experimental system.

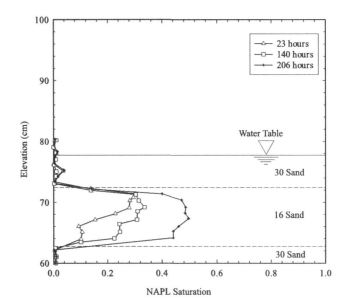

Figure 3.27. Saturation profiles showing a LNAPL preferentially flowing in a more permeable layer.

proceeded to migrate preferentially along the coarse layer in the direction of groundwater flow. Visual observations showed that the upper interface between the two sands confined the NAPL flow within the coarse layer. The buoyancy forces of the LNAPL were insufficient to overcome the entry pressure and to push the LNAPL into the finer sand. The LNAPL remained permanently entrapped below the water table within the coarse sand, with saturations significantly higher (estimated as 80%) than residual levels in the #16 sand (estimated as 20-30%).

A scenario where a DNAPL encounters a coarse layer of material is shown in Figure 3.28. As discussed earlier, the DNAPL will move downward from the source location and penetrate the water table. If the conditions are appropriate, instability at DNAPL/water interface will develop and the DNAPL will flow through fingers. When the fingers encounter the interface between fine and coarse sand, they will enter the coarser material. After entering the coarse layer, the DNAPL will accumulate at the interface between the coarse (soil 2) and fine (soil 1) due to the barrier effects created by the high displacement pressure of the finer soil. However, even if the macroscopic conditions have not been met for the DNAPL to enter the fine soil as a front, it is possible for the DNAPL to move across the coarse/fine soil interface through fingering. This phenomenon was observed by Illangasekare et al. [1995b] in spill simulations conducted in the laboratory.

To obtain an understanding of the effect of a heterogeneous field on the migration and entrapment of DNAPLs, a set of experiments were conducted in soil tanks packed to represent a randomly heterogeneous hydraulic conductivity field [Compos, 1998]. Five well characterized laboratory test sands (#8, #16, #30, #70

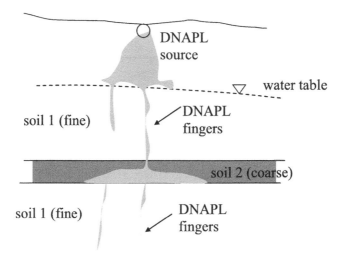

Figure 3.28. DNAPLs in high permeability formations.

and #110) were used. These sands represent a range of values of hydraulic
conductivity (from 1.20 cm/s to 0.004 cm/s) and capillary pressure vs. saturation
characteristics. The tanks were packed in layers that were 10 cm long and 2 cm
thick to obtain a variance in ln(K) of 2.86 and horizontal and vertical correlation
lengths of 20 cm and 4 cm, respectively. Five different packings with identical
statistics but random distributions of the test sands were prepared.

The DNAPL saturation distribution was observed visually and recorded using
a dual gamma system. Dual gamma systems are used to indirectly measure fluid
saturations by placing two different gamma radiation sources on one side of an
experimental container and measuring the amount of gamma radiation that penetrates to the other side of the container. The difference between the source radiation and radiation measured on the other side of the container depends on the
fluid and solid contents in the porous media-fluid mixture. By using multiple
gamma sources with different wavelengths and assuming that each fluid atte-
nuates the radiation to different extents, multiple fluid (i.e. DNAPL and water)
saturations can be determined.

The DNAPL (trichloroethane, or TCA) was introduced directly below the water
table at a rate of 50 ml/min for 54 min duration. The water table was located at
an elevation 70 cm above the base of the tank. The DNAPL migration pattern
consisted of spreading at interfaces due to the capillary barrier effect, preferential
flow through coarser layers, and fingering. The distribution of the DNAPL
90 minutes after the release began is shown in Figure 3.29. The results show that
the random heterogeneity resulted in an almost random entrapment distribution
of the DNAPL. The normalized distribution of DNAPL in the five soils is shown
in Figure 3.30 for the five different packings. The high normalized NAPL percentage within the higher permeability sands further demonstrates that free phase
DNAPLs are preferentially entrapped in high permeability regions of aquifers.

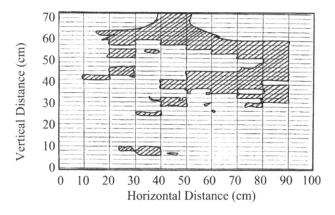

Figure 3.29. DNAPL entrapment in a random field.

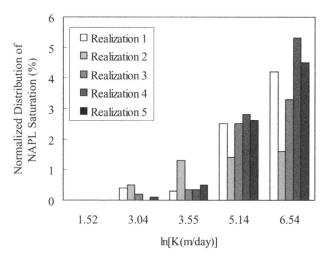

Figure 3.30. Normalized distribution of NAPL saturation in five porous medium for five different realizations.

3.5   SMALL SCALE NAPL DISTRIBUTION

3.5.1   Unstable Fronts and Fingers

Summary: DNAPLs after migrating through the unsaturated zone penetrate the saturated zone by first displacing the water in the capillary fringe that is under suction. As the DNAPL densities are higher than that of the water, assisted by gravity, the DNAPL displaces the water in the saturated soil pores. Depending on properties of the fluid and soil, unstable fronts can develop at the macroscopic DNAPL/water interface. These fronts initiate fingers that carry DNAPLs preferentially through the saturated zone. The unstable behavior produces complex entrapment configurations in heterogeneous aquifers.

The final DNAPL distribution in the spill experiment discussed in the previous section (Figure 3.29) shows that the DNAPL moved preferentially through vertical channels created as a result of unstable displacement of water by the DNAPL. This phenomenon is suspected to be the mechanism by which DNAPLs move rapidly through aquifers and reach the bedrock, where they pool. During the vertical propagation, these fingered channels may encounter heterogeneities (e.g. capillary barriers) and form suspended DNAPL pools in the formation. As the factors that control finger initiation are difficult to quantify, it is not possible to predict where the instabilities occur and hence the finger pathways. Fortunately,

in almost all DNAPL remediation problems, it is not necessary to make such predictions. However, it will be of practical value to be able to evaluate under what
conditions and when it is possible for the fingers to form to assist in determining
possible final entrapment DNAPL entrapment architecture. For example, with a
fundamental understanding of how unstable fingers are formed in combination
with specific site conditions such as spill volume and history, spill area, type of
DNAPL and geology of the formation it may be possible to develop conceptual
models of DNAPL migration and entrapment. Such knowledge will help in developing effective site characterization techniques to determine entrapment architecture that will be of used in developing DNAPL source zone treatment and cleanup
strategies. The goal of this section is to provide such a theoretical foundation on
unstable fingering of DNAPLs in saturated soils.

The factors that influence the stability of immiscible displacement of fluids in
porous media have been identified in the literature on petroleum engineering and
theoretical physics (e.g. Chuoke et al. [1959], Lenormand [1985]). These factors
include: (1) relative viscosities of the displaced and displacing fluids (mobility),
(2) gravity, (3) capillary forces, (4) system permeability, (5) wettability and contact angle, (6) displacement velocity, and (7) system geometry and dimensions.
At the pore scale, fingering is initiated as a result of small perturbations that are
amplified as a result of instabilities. These instabilities are created by a combination of factors that depend on the properties of the displacing and displaced fluids,
pore characteristics and a critical wavelength of the perturbation.

We can describe a scenario where a non-wetting fluid displaces wetting fluid
vertically downward through a homogenous and isotropic porous medium.
Figure 3.31(a) shows the sharp DNAPL front that forms initially at the pore scale.
The front as shown is stable because all the microscopic DNAPL fronts are at the
same elevation, thus creating a sharp front at the macroscopic scale. The macroscopic interface (or the zone separating the two fluids) is subject to a slight
perturbation at the pore scale as shown in Figure 3.31(b).

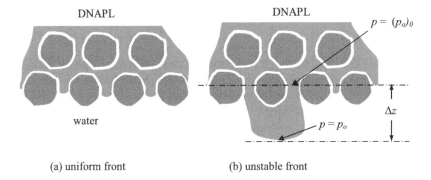

(a) uniform front   (b) unstable front

Figure 3.31. Instability at DNAPL/water front.

Assume that the perturbation has moved to depth Δz and is propagating with a velocity v. The dynamics of the moving boundary can then be described using the pressure difference across the microscopic interface. Let $(p_o)_0$ be the pressure within the DNAPL at the macroscopic interface. Applying Darcy's law for the unstable fluid flow at the macroscopic scale:

$$v = -\frac{k_o}{\mu_o} \frac{p_o - [(p_o)_0 + \Delta z \rho_o g]}{\Delta z} \tag{3.25}$$

where p_o is the pressure in the DNAPL phase behind the front, ρ_o is the density of the DNAPL, k_o is the DNAPL permeability, μ_o is the dynamic viscosity of the DNAPL, g is the gravitational acceleration and Δz is the vertical dislocation of the macroscopic front. From (3.25) an expression can be derived for the pressure in the DNAPL at the macroscopic interface:

$$p_o = (p_o)_0 + \rho_o g \Delta z - \frac{\mu_o v \Delta z}{k_o} \tag{3.26}$$

By definition, the capillary pressure at the DNAPL/water interface is given by:

$$p_c = p_o - p_w \tag{3.27}$$

Writing Equation (3.25) for the water Darcy velocities on the water side of the interface, we obtain

$$p_w = (p_o)_0 + \rho_w g \Delta z - \frac{\mu_w v \Delta z}{k_w} - p_c \tag{3.28}$$

where k_w is the permeability of water, ρ_w is the density of water and μ_w is the dynamic viscosity of water. Substituting Equations (3.26) and (3.28) in (3.27) and rearranging yields a critical displacement velocity of

$$v_{crit} = \frac{g(\rho_w - \rho_o)}{\dfrac{\mu_w}{k_w} - \dfrac{\mu_o}{k_o}} \tag{3.29}$$

The stability conditions of the perturbation can be analyzed using the expression for the critical velocity. Note that this expression contains properties of the porous medium as well properties of the two fluids. Because the DNAPL is heavier than water ($\rho_o > \rho_w$), and for a DNAPL that is less viscous than water ($\mu_o < \mu_w$), one finds from Equation (3.29), $v_{crit} < 0$, even at $v = 0$ (where v is the macroscopic velocity). In this case, both gravity and viscosity configurations act to destabilize the DNAPL/water interface at the macroscopic scale. If the DNAPL is more viscous than water ($\mu_o > \mu_w$), from Equation (3.29) it can be seen that a possible condition can occur (depending on relative values of k_0 and k_w) where $v_{crit} > 0$, That is, unstable displacement occurs at $v < v_{crit}$ for slow displacement.

The above discussion suggested by Hill [1952] is a simplification of complex dynamics that occur in finger formation, yet it helps to illustrate the basic concept. It is more appropriate to evaluate the phase mobilities at the front than the viscosity differences. In addition, surface tension affects the displacement process and may not stabilize flow that is determined to be stable using the criterion discussed above. The stability analysis can be performed by assuming a disturbance and determine the growth or decay of the disturbance. The initial disturbance can be represented using a function that is defined using a wavelength, λ. A critical value for this wavelength can then be determined. When the wavelength exceeds this critical value, the displacement will grow resulting in the initiation of a finger.

Chuoke et al. [1959] derived a critical wavelength λ_{crit} based on the critical velocity (Equation (3.29)), applied to Hele-Shaw cells. A Hele-Shaw cell consists of a pair of transparent plates separated by a small gap, typically 0.5 mm. In a typical Hele-Shaw cell experiment, a fluid that initially occupies the cell is displaced by a second, immiscible fluid. The transparency of the cell walls allows for direct observation of finger initiation and measurement of finger wavelength, speed, and dimensions. The expression for critical wavelength derived by Chuoke et al. [1959] is

$$\lambda_{crit} = 2\pi \left[\frac{\sigma^*}{\left(\dfrac{\mu_w}{k_w} - \dfrac{\mu_o}{k_o} \right)(v - v_{crit})} \right]^{1/2} \tag{3.30}$$

where σ^* is the macroscopic effective surface tension.

In extending the theories derived for simple systems such as Hele-Shaw cells to porous media, it is necessary to recognize that many pore-scale mechanisms complicate the flow in porous media and unstable displacement [Homsey, 1987]. These mechanisms include details of the wetting behavior, dynamics of the wetting interface and contact angles, static stabilities of interface curvatures supporting the fluid head, dynamic instabilities of blobs of immiscible non-wetting fluid, and mass transfer across interfaces [Held, 1993].

Lenormand [1985] proposed a phase diagram (Figure 3.32) to incorporate some of these factors and to distinguish between different fingering regimes. Lenormand [1985] conducted two-dimensional experiments to identify three types of interface morphologies that depend on the viscosity ratio, M, of the two fluids and the capillary number, C. These two parameters appear in Figure 3.32 and are defined as:

$$M = \frac{\mu_2}{\mu_1} \tag{3.31}$$

and

$$C = \frac{v\mu_2}{A\sigma\cos\theta} \tag{3.32}$$

where μ_1 is the viscosity of the displaced fluid and μ_2 is the viscosity of the displacing fluid, θ is the fluid-solid contact angle, A is the cross-sectional area of the

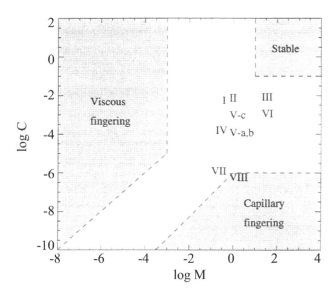

Figure 3.32. Phase-diagram of displacement regimes [after Lenmorand, 1985].

sample, and v/A is the average interstitial velocity. In the problem that is of interest to us, the displacing fluid is the DNAPL and the displaced fluid is water. The three regimes identified in the Figure 3.32 are: (a) stable displacement regime, (b) viscous fingering regime and (c) capillary fingering regime. According to Homsey [1987], these results can only be considered qualitative and specific to the apparatus used. Nevertheless, they are helpful to distinguish fingering regimes.

Held and Illangasekare [1995a] conducted a set of experiments where three DNAPLs (TCE, TCA and dibutyl phthalate) were spilled in a three-dimensional tank homogeneously packed with three test sands. The properties of the three test DNAPLs and the three sands are given in Tables 3.3 and 3.4. A total of eight experiments were conducted. The viscosity ratios (M) and the capillary numbers (C) associated with these experiments placed them in the locations that are identified on the phase diagram (Figure 3.32). The data show that the experimental conditions used fall outside the stable, viscous and capillary finger regimes (unshaded area). This suggests that for the three common types of DNAPLs used and for the range of soils, the observed fingers are a result of combined capillary and gravity effects. The results presented are qualitative and depend on the configuration of the test system. The ability to represent the laboratory test conditions suggests that the parameters M and C could be used to describe finger regimes in practical situations. However, phase-diagrams such as in Figure 3.32 need to be developed to cover the diversity of conditions that occur in the field [Held and Illangasekare 1995b].

The question of validity of the use of continuum-based flow and transport equations to describe unstable displacement of fingering is subjected to considerable

TABLE 3.3.   Physical properties of DNAPLs used in fingering experiments of Held and Illangasekare [1995a] (measured at 20°C).

DNAPL	Density (g/cm^3)	Viscosity (cps)	Surface Tension (dyne/cm)
Trichloroethylene TCE	1.460	0.57	35
1,1,1-Trichloroethane TCA	1.350	0.84	45
Dibutyl phthalate	1.044	20	35

TABLE 3.4.   Properties of porous media used in finger experiments of Held and Illangasekare [1995a].

Mesh Size of Porous Media	Mean grain size d_{50} (mm)	Porosity	Hydraulic Conductivity (cm/s)	Brooks-Corey p_d (mm of water)	Brooks-Corey λ
#8	1.468	0.427	1.20	37	4.8
#30	0.436	0.440	0.15	55	2.1
#70	0.198	0.425	0.02	250	1.9

skepticism. Flow in porous media is conventionally treated by averaging microscopic parameters over some representative elementary volume. Since there are many unknowns in the formulation of the problem, DNAPL fingering has been omitted in predictive modeling. Ignoring the fingering in modeling may tend to underestimate the depth of migration of DNAPLs, as fingering may produce preferential channels for the DNAPLs. Fingering also is significant in that it can provide pathways for DNAPLs to migrate from pools that are trapped, in a macroscopic sense, on top of low permeability materials.

3.5.2   Occurrence of NAPLs in Clay or Rock Fractures

Summary: To study the occurrence and behavior of nonaqueous phase liquids (NAPLs) in fractured materials, one needs to focus on the following three questions. (a) What are the conditions that allow entrance of NAPLs into a given fracture? (b) How far can NAPLs penetrate through the fracture? (c) What happens to NAPLs in any given fracture?

In Figure 3.33, we illustrate a fractured rock such as basalt or limestone with fractures extending all the way to the ground surface. The water table is present at a certain depth below the ground surface. The fractures have different apertures and could be open, partially open or completely filled with clay or other materials. Depending on the fracture aperture and the degree of its filling, the water level in the fracture will rise to above the water table, due to capillary effects. For

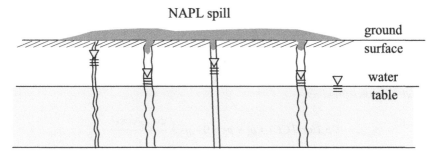

Figure 3.33. Schematic representation of DNAPL penetrating a fracture system.

an open fracture, the amount of water level rise is inversely proportional to the fracture aperture. If we have a release of NAPL on the ground surface, the NAPL will enter all open fractures, independent of their apertures, and move downward until it reaches the saturated zone, where fractures are completely filled with water. From here on, the movement will continue until capillary pressure becomes equal to the entry pressure, p_d [Kueper and McWhorter, 1991]. The entry pressure for two parallel-plates fractures may be expressed by:

$$p_d = \frac{2\sigma\cos\theta}{e} \qquad (3.33)$$

where e is the fracture aperture, σ is the interfacial tension between NAPL and water and θ is the contact angle measured through the wetting phase.

Example: Consider an open fracture with an aperture of 0.1 mm. Assume that the surface tension of water and TCE are 72.8 and 29.3 dyne/cm, respectively, the interfacial tension between water and TCE is 34.5 dyne/cm, and the contact angle θ is zero. (a) Calculate the capillary rise of water in this fracture. (b) If a column of 5 cm high of TCE is introduced into this fracture, estimate the depth of TCE penetration into the water column. (c) What would be the depth of penetration if we introduce a column of 5 cm of benzene?

Solution: (a) Calculate the capillary rise of water in this fracture:

The capillary rise of water may be calculated from

$$h_w = \frac{2\sigma\cos\theta}{e\cdot\gamma_w}$$

where σ and γ_w are surface tension and unit weight of water, respectively. Therefore,

$$h_w = \frac{2\times 72.8\,\text{dyne}/\text{cm}}{0.01\,\text{cm}\times 1{,}000\,\text{dyne}/\text{cm}^3} = 14.56\,\text{cm}$$

(b) If a column of 5 cm high of TCE is introduced into this fracture, estimate the depth of TCE penetration into the water column. Let us consider four points as shown in Figure 3.34. Points 1 and 2 are located immediately above and below the interface of air and TCE. Points 3 and 4 are immediately above and below the TCE and water interface. The capillary pressure between air and TCE is defined as:

$$p_c(air, TCE) = p_1 - p_2 = 0 - p_2 = \frac{2\sigma_{a,T}\cos\theta}{e}$$

$$p_2 = -\frac{2 \times 29.3 \text{ dyne/cm}}{0.01 \text{ cm}} = -5,860 \text{ dyne/cm}^2$$

$$p_3 = p_2 + h_{TCE} \cdot \gamma_{TCE} = -5,860 + (5 \text{ cm} \times 1,460 \text{ dyne/cm}^3)$$
$$p_3 = 1,440 \text{ dyne/cm}^2$$

At equilibrium, the capillary pressure between TCE and water will be equal to the entry pressure:

$$p_c(TCE, water) = p_3 - p_4 = \frac{2\sigma_{T,w}\cos\theta}{e}$$

$$p_4 = p_3 - \frac{2 \times 34.5 \text{ dyne/cm}}{0.01 \text{ cm}} = 1,440 - 6,900 = -5,460 \text{ dyne/cm}^2$$

The corresponding capillary rise of water is:

$$h_w = \frac{5,460 \text{ dyne/cm}^2}{1,000 \text{ dyne/cm}^3} = 5.46 \text{ cm}$$

The depth of TCE penetration is $h_w - h'_w = 14.56 - 5.46 = 9.1$ cm.

Another way to look at this problem is to write an equation for balance of forces in the fracture:

$$h_w\gamma_w + h_{TCE} \cdot \gamma_{TCE} = p_c(air, TCE) + p_c(water, TCE)$$

$$h_w = \frac{[(5,860 + 6,900) - (5 \times 1,460)] \text{ dyne/cm}^2}{1,000 \text{ dyne/cm}^3}$$

$$h_w = \frac{12,760 - 7,300 \text{ dyne/cm}^2}{1,000 \text{ dyne/cm}^3} = 5.46 \text{ cm}$$

(c) What would be the depth of penetration if we introduce a column of 5 cm of benzene?

The surface tension of benzene and interfacial tension between water and benzene are 28.9 and 35 dyne/cm, respectively. Once again, if we apply the equation of equilibrium of forces:

$$h_w\gamma_w + h_B\gamma_B = p_c(air,\ B) + p_c(water,\ B)$$

$$p_c(air,B) = \frac{2\sigma_{A,B}\cos\theta}{e} = \frac{2\times 28.9\ \text{dyne}/\text{cm}}{0.01\ \text{cm}} = 5,780\ \text{dyne}/\text{cm}^2$$

$$p_c(water,B) = \frac{2\sigma_{W,B}\cos\theta}{e} = \frac{2\times 35\ \text{dyne}/\text{cm}}{0.01\ \text{cm}} = 7,000\ \text{dyne}/\text{cm}^2$$

$$h_w = \frac{[(5,780+7,000)-(5\times 870)]\ \text{dyne}/\text{cm}^2}{1,000\ \text{dyne}/\text{cm}^3} = 8.43\ \text{cm}$$

In Figure 3.35, a fractured aquitard separates two high permeability porous media aquifers. The top aquifer is unconfined. If a DNAPL release occurs underground in the vadose zone and conditions allow, it may migrate down to the water table. If the aquifer is permeable enough and the volume of release is large enough, then DNAPL may migrate to the bottom of the top aquifer. In this case, the fractures in the aquitard are likely to be completely saturated with water. The mechanism of DNAPL entry into saturated fractures is discussed by Kueper and McWhorter [1991]. Here, in order for DNAPL to enter an open fracture, the capillary pressure, p_c, at the entrance to the fracture must exceed the entry pressure, p_d, of the fracture. Since, as defined in Equation (3.33), p_d is inversely proportional to the fracture aperture, a significant DNAPL head may need to accumulate before the DNAPL can enter small aperture fracture. Again, once

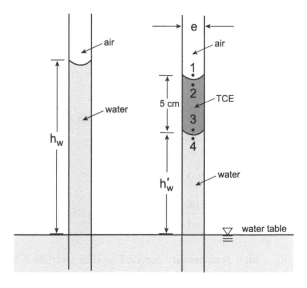

Figure 3.34. Schematic diagram of water and TCE in open fractures.

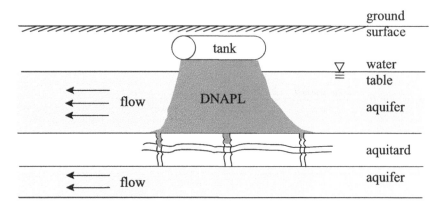

Figure 3.35. Schematic representation of DNAPL penetrating a fractured aquitard.

DNAPL enters the fracture, it will move downward until it reaches a point where $p_c = p_d$. Given a sufficient accumulation of DNAPL head, the DNAPL may eventually arrive at the lower aquifer.

Example: Estimate the required height of a TCE pool for entry into a fracture in the aquitard with an aperture of 0.01 mm.

Solution: The capillary pressure at the top of the fracture may be given by

$$p_c = h(\gamma_{TCE} - \gamma_w)$$

where h is the height of TCE pool at the bottom of the top aquifer. The condition for the entry of TCE into the fracture is that p_c should be equal or greater than the entry pressure p_d.

$$p_d = \frac{2\sigma_{w,T}\cos\theta}{e}$$

$$p_d = \frac{2 \times 34.5 \text{ dyne/cm}}{0.001 \text{ cm}} = 69,000 \text{ dyne/cm}^2$$

The inequality $p_c \geq p_d$

requires that $h(\gamma_{TCE} - \gamma_w) \geq 69,000$ dyne/cm^2

or

$$h \geq \frac{69,000 \text{ dyne/cm}^2}{(1,460 - 1,000) \text{ dyne/cm}^3} = 150 \text{ cm}$$

In Figure 3.36, a high permeability perched aquifer overlies a fractured clay layer. The porous material beneath the clay layer is not saturated. Another unconfined aquifer is present at a lower elevation. Here, although fractures could be

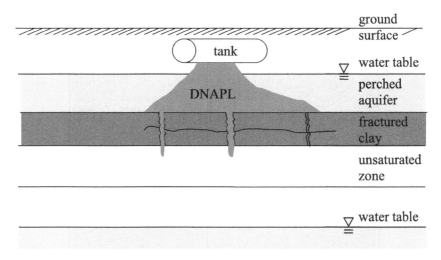

Figure 3.36. Schematic representation of DNAPL penetrating a perched aquifer system.

saturated with water, if a DNAPL arrives at the top of the fractures in the clay, it will penetrate the fractures and migrate to the vadose zone beneath the clay layer [Stephens et al., 1998].

None of the three cases discussed above account for dissolution of NAPLs into the water layer that may be present between the NAPL and fracture wall or for molecular diffusion of dissolved chemicals into the rock matrix. Parker et al. [1994] presented a modified conceptual model that considers such processes. In this model, the NAPL is mobile only in the fractures. The rock matrix contains immobile water. The fracture may also contain a film of water between the NAPL and the fracture wall (see Figure 3.37). Through dissolution and diffusion into the matrix, a small volume of NAPL release in the fractured medium can spread extensively and occupy a relatively large volume of the bulk medium, even if the void space of the fractures constitutes a small portion of the total rock bulk volume.

Dissolution of NAPL into the water in the fracture creates a concentration gradient in the aqueous phase directed from the fractures into the matrix. As a result, molecules of chemicals will move from the high concentration zone in the fracture into the lower concentration zone in the rock matrix, allowing further dissolution of NAPL in the water in the fracture. The extent of solute penetration into the matrix depends on the NAPL volume available in the fracture and the time that NAPL is available, along with the NAPL solubility in the water, adsorption onto the matrix material, and the tortuosity of the matrix material. If the matrix blocks are relatively small and release of NAPLs continues over a long period of time, dissolved chemicals with concentrations close to their solubility will extend throughout the total matrix block.

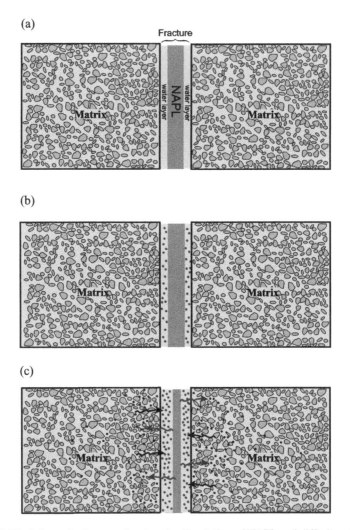

Figure 3.37. Schematic diagram showing the dissolution of NAPL and diffusion of solute into the rock matrix: (a) NAPL enters fracture, (b) NAPL dissolves into water layer, (c) dissolved NAPL diffuses into matrix.

In Figure 3.38, we show an idealized system with vertically-oriented fractures that are spaced equally through the matrix. In this case, we can describe the transport of dissolved NAPL into the matrix as a one-dimensional diffusion process. We further idealize the system by saying that the NAPL is instantaneously placed in the fracture, the water surrounding the NAPL in the fracture is instantaneously in equilibrium with the NAPL. Then, the governing equation, initial condition,

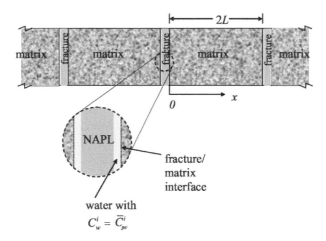

Figure 3.38. Schematic of idealized fracture/matrix system where the fracture is contaminated with NAPL.

and boundary conditions for transport of dissolved NAPL in the matrix are

$$\frac{\partial C_w^i}{\partial t} = D_w^{i,eff} \frac{\partial^2 C_w^i}{\partial x^2}$$

$$C_w^i(x, t < 0) = 0$$

$$C_w^i(x = 0, t \geq 0) = \bar{C}_w^i$$

$$\left.\frac{\partial C_w^i}{\partial x}\right|_{x=L} = 0, \quad for\, t \geq 0 \tag{3.34}$$

where $D_w^{i,eff} = \phi^{1.33} D_w^i$. The initial condition indicates that the concentration of dissolved NAPL in the matrix is initially zero. The boundary conditions at $x = 0$ and $x = L$ indicate that the water surrounding the NAPL in the fracture is in equilibrium with the NAPL and that there is symmetry at the midpoint of the matrix, respectively. The solution to Equations (3.34) is [Carslaw and Jaeger, 1988].

$$\frac{C_w^i}{\bar{C}_w^i} = 1 + \frac{2}{L}\sum_{n=0}^{\infty} \exp\left[-D_w^{i,eff}(2n+1)^2 \pi^2 t / 4L^2\right]\cos\left[\frac{(2n+1)\pi x}{2L}\right]\frac{2L(-1)^{n+1}}{(2n+1)\pi} \tag{3.35}$$

Example: Calculate the concentration of dissolved TCE in a matrix as a function of distance and time for a fracture/matrix system described in Figure 3.38, given $2L = 2\ m$, $\phi = 0.1$, and $D_w^{i*} = 8.8 \times 10^{-10}\ m^2/s$ for TCE.

Solution: In Figure 3.39, the concentration in the matrix as a function of distance from the fracture/matrix interface is shown for 5, 50 and 500 years after the diffusion into matrix began (t ≥ 0). The results show that, only after about 500 years, the concentration of dissolved NAPL is near the solubility level throughout the matrix.

A spreadsheet file (NAPL_matrix_diffusion.xls) *that can be used to make calculations of TCE transport in a matrix from a NAPL source in a fracture is found on the accompanying CD.*

In the previous discussions, we idealized the walls of the fracture as smooth; that is, the aperture of the fracture is constant. In reality, there is a distribution of apertures with the fracture, so that the fracture is not smooth. We also assumed that that the NAPL would completely saturate the fracture. In most situations, however, we will see a drainage-imbibition process similar to that observed in granular media. The NAPL will initially invade the fracture, displacing a fraction of the resident water and reaching a maximum NAPL saturation (water drainage). The NAPL will not completely displace the resident water, if we have the case where the water wets the wall of the fracture. The magnitude of the maximum NAPL saturation is controlled by the aperture distribution and the pressure driving the NAPL into the fracture.

When the supply of NAPL is exhausted or cut off from the fracture, the NAPL will release from the fracture until a minimum, or residual, NAPL saturation is achieved. Again, aperture distribution will determine the magnitude of the

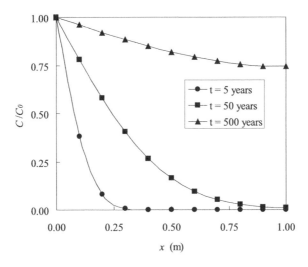

Figure 3.39. Concentration in matrix as a function of distance from the fracture/matrix interface ($x = 0$) at various times.

residual NAPL saturation. Two movie files (TCE_flood.mov and water_flood.mov), found on the accompanying CD, illustrate the invasion and release of a NAPL in a transparent, experimental fracture. The behavior of the NAPL in the invasion and release steps demonstrate the impact of the aperture distribution on the NAPL movement and eventual, residual NAPL distribution. Refer to the file fracture_movie_notes on the accompanying CD, for an explanation of the experimental system.

4

Site Characterization and Monitoring

Determination of the presence of NAPLs and their distribution at contaminated sites is critical for determining the risks associated with NAPL contamination and for determining the need for and potential design of site remediation efforts. In the following chapter, we describe common observations associated with monitoring and assessment of sites contaminated with NAPLs. In particular, we focus on observations of NAPLs in monitoring wells and soil samples (Section 4.1) and observations of NAPLs dissolved in groundwater (Section 4.2). Advanced technologies for characterizing NAPLs at contaminated sites include inter-well partitioning tracer tests (see, e.g., Jin et al. [1995]), surface geophysical techniques (http://enviro.nfesc.navy.mil/erb/restoration/technologies/invest/geo_phys/gp-01.asp), and light-based techniques associated with direct-push technologies (http://enviro.nfesc.navy.mil/erb/restoration/technologies/invest/geo_phys/gp-05cpt.htm). Many of these technologies were reviewed by Mercer and Cohen [1990] and Kram et al. [2001, 2002].

4.1   LNAPL OBSERVATIONS

4.1.1   Discrepancy Between Free Product Levels in Monitoring Wells and LNAPL Specific Volume

Summary: Observations of LNAPL thickness from monitoring wells can dramatically over- or under-estimate the actual LNAPL thickness and volumes in aquifers. Discrepancies between well observations and LNAPL thickness and volume in soils are the result of porous medium retention characteristics, saturation history, and hysteretic behavior. In this section, correlations between LNAPL well thickness, soil hydrocarbon thickness (depth over which the LNAPL saturation is > 0), and LNAPL specific volume (volume in the soil per unit area) are derived and illustrated.

Surface spills and leakage from underground storage tanks of NAPLs are a widespread source of groundwater contamination. Light nonaqueous phase

Soil and Groundwater Contamination: Nonaqueous Phase Liquids
Water Resources Monograph 17
Copyright 2005 by the American Geophysical Union
10.1029/17WM04

liquids (LNAPLs) may accumulate above the water-saturated regions of the sub-surface and serve as a source of volatile and/or dissolved components. The distribution of LNAPL is a function of water, LNAPL, and air pressures, as well as fluid and porous medium properties.

To assess spill volumes and to design and monitor remediation operations, observation wells are commonly installed to measure LNAPL thickness. Unfortunately, the interpretation of observed LNAPL thickness data has some inherent difficulties. It has been known for a long time that the actual LNAPL volume per unit surface area (LNAPL specific volume) is less than the LNAPL thickness in a well [van Dam, 1967]. After a review of numerous case histories at LNAPL sites, Marinelli and Durnford [1996] listed some common, apparent discrepancies:

- Monitoring wells contain no LNAPL even though soil sampling indicates presence of considerable amount of LNAPL in the adjacent formation in the vadose zone or below the water table.
- LNAPL thickness in a well increases when the water table falls and decreases when the water table rises.
- Sudden appearances or disappearance of observable LNAPL in wells.
- LNAPL completely disappears from monitoring wells when the water table level drops below a historic minimum.

A field site with fluctuating LNAPL thicknesses in monitoring wells has been described by Kemblowski and Chiang [1990]. An example of observations for one of the wells is shown in Figure 4.1. The picture demonstrates the time-variant behavior mentioned by Marinelli and Durnford [1996]: a drop in the observed LNAPL well thickness from 15 to 4 ft was initially observed when the water-LNAPL interface elevation above a reference level was increased from 5 to 18 ft, while later on the observed LNAPL well thickness increased from 4 to 18 ft when the water- LNAPL interface dropped from 18 to 0 ft.

Several authors have proposed simple conversions from measured LNAPL thickness in monitoring well to LNAPL specific volumes (see e.g., Pastrovich et al. [1967], Hall et al. [1984]) that do not consider porous medium properties. Lenhard and Parker [1990] and Farr et al. [1990] were the first to propose physically-based methods that, under the assumption of vertical equilibrium, predict water and LNAPL saturation profiles in porous media. Residual saturation in the vadose zone and LNAPL entrapment by water are not explicitly accounted for in their theories. If fluid pressure distributions in the subsurface can be inferred from well fluid levels, and water-LNAPL-air saturation-capillary pressure relations for the soil are known, fluid saturation distributions can be predicted and integrated to compute the LNAPL hydrocarbon volume. An understanding of these methods is important to appreciate noted discrepancies between LNAPL well thickness and LNAPL thickness in the formation.

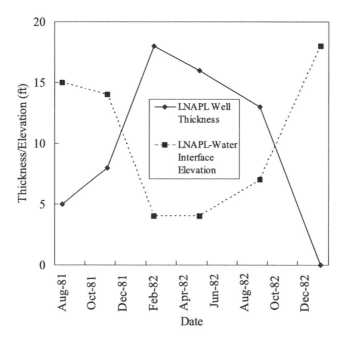

Figure 4.1. Fluctuation of the LNAPL-water interface elevation and LNAPL well thickness at a field site (adapted from Kemblowski and Chiang [1990]; reprinted with permission from the National Water Well Association, Dublin, OH).

The discussion starts by defining water-height-equivalent pressure heads of water and LNAPL as:

$$h_w = p_w/g\rho_w \qquad (4.1)$$

$$h_o = p_o/g\rho_w \qquad (4.2)$$

where p_w and p_o are water and LNAPL phase pressures, g is the gravitational acceleration and ρ_w is the density of water. Consider a system with air, LNAPL, and water in which a screened well and a piezometer are installed according to Figure 4.2. LNAPL is observed in the well, characterized by an air-LNAPL table elevation z_{ao}, at which the gauge LNAPL pressure is zero, and the LNAPL-water table elevation, z_{ow}, where the water and LNAPL pressures are equal. From the piezometer, an air-water table elevation, z_{aw}, can be defined where the gauge water pressure is zero. Assuming an LNAPL specific gravity (ratio of LNAPL to water density) of ρ_{ro}, the water and LNAPL pressure heads at an elevation z can be written as

$$h_w(z) = z_{aw} - z \qquad (4.3)$$

$$h_o(z) = \rho_{ro}(z_{ao} - z) \qquad (4.4)$$

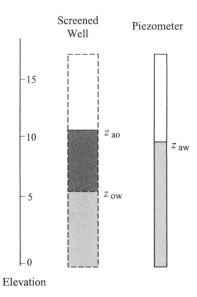

Figure 4.2. Fluid levels in monitoring wells.

The relation between the two LNAPL-water interface elevations can be found by using the fact that $p_o = p_w$ (thus $h_o = h_w$) at $z = z_{ow}$. Thus, one obtains:

$$z_{aw} = (1 - \rho_{ro})z_{ow} + \rho_{ro}z_{ao} \qquad (4.5)$$

From Equation (4.5) it is clear that knowledge of any two out of the three fluid table levels is sufficient to define water-LNAPL-air static vertical head distributions. For the example shown in Figure 4.2, where the LNAPL has a ρ_{ro} of 0.73 and z_{ow} and z_{ao} are, respectively, 5 and 10 m above a certain reference level, the water table elevation $z_{aw} = [(1 - 0.73) \times 5] + (0.73 \times 10) = 8.65$ m. The equation also implies that, if a monitoring well exists, installation of a piezometer is not required.

Because fluid saturations depend directly on pressure differences between phases, the following capillary heads are introduced

$$h_{ao} = h_a - h_o \qquad (4.6)$$

$$h_{ow} = h_o - h_w \qquad (4.7)$$

Assuming $h_a = 0$, the following expressions for h_{ao} and h_{ow} as a function of elevation may be derived from (4.5)–(4.7):

$$h_{ao} = \rho_{ro}(z - z_{ao}) \qquad (4.8)$$

$$h_{ow} = (1 - \rho_{ro})(z - z_{ow}) \qquad (4.9)$$

Equations (4.8) and (4.9) show that the capillary head for fluid pair i-j only depends on the distance to the i-j table and the relative density.

To describe vertical fluid saturation distributions, saturation-capillary head relations must be known. Adopting the Leverett assumption [Leverett, 1941] that in a water-wet porous medium, effective water saturation and total liquid (water + LNAPL) saturation are functions of the water-LNAPL and LNAPL-air capillary heads, respectively, Parker et al. [1987] proposed that these functions may be scaled according to

$$S_{ew} = S^*(\beta_{ow}h_{ow}) \qquad (4.10)$$

$$S_{et} = S^*(\beta_{ao}h_{ao}) \qquad (4.11)$$

where $S^*(h^*)$ is the scaled capillary head relation defined below, $S_{ew} = (S_w - S_{wir})/(1 - S_{wir})$ and $S_{et} = (S_w + S_o - S_{wir})/(1 - S_{wir})$. In Equations (4.10) and (4.11), S_w and S_o are actual water and LNAPL saturations, and S_{wir} represents the irreducible water saturation. The scaling factors β_{ow} and β_{ao} are fluid-pair dependent and are usually estimated from water-LNAPL and air-LNAPL interfacial tension (σ_{ij}) data using $\beta_{ao} = \sigma_{aw}/\sigma_{ao}$ and $\beta_{ow} = \sigma_{aw}/\sigma_{ow}$.

Given an appropriate expression for the scaled saturation-capillary head relation, Equations (4.8) and (4.9) can be inserted in Equations (4.10) and (4.11) to determine vertical saturation distributions. For two- and three-phase conditions, the van Genuchten [1980] and Brooks and Corey [1964] relations are widely used.

With Equations (4.1)–(4.11), the van Genuchten model (Equation (2.8)), and the Brooks and Corey model, (Equation (2.6)), the necessary tools are available to determine relations for the soil hydrocarbon thickness D_o, and the soil specific volume, V_o. These are defined as the depth over which $S_o > 0$, and the LNAPL volume in the soil per unit area in the horizontal plane, respectively. Both parameters have been used in the literature, often without proper definitions. As a result, there is considerable confusion and even misuse of these parameters.

To compute the soil hydrocarbon thickness, D_o, we need to determine the upper and lower limit of the zone where $S_o > 0$. The upper limit, z_{ou}, i.e., the vertical location were the water saturation is equal to the total liquid saturation can be calculated by equating Equations (4.10) and (4.11). That is, for $\overline{S}_w$ and $\overline{S}_t$ to be equal, $\beta_{ow}h_{ow}$ is set equal to $\beta_{ao}h_{ao}$ at $z = z_{ou}$. Employing the definitions of the capillary heads given in Equations (4.8) and (4.9), the upper limit is given by

$$z_{ou} = \frac{z_{ao}\beta_{ao}\rho_{ro} - z_{ow}\beta_{ow}(1 - \rho_{ro})}{\beta_{ao}\rho_{ro} - \beta_{ow}(1 - \rho_{ro})} \qquad (4.12)$$

For van Genuchten S-h relations, where the LNAPL fluid entry head equals zero, the lower limit, z_{ol}, is equal to z_{ow}. Since $D_o = z_{ou} - z_{ol}$, the soil hydrocarbon thickness applicable to van Genuchten relations can be written as:

$$D_o = z_{ou} - z_{ol} = \frac{\rho_{ro}\beta_{ao}H_o}{\rho_{ro}\beta_{ao} - \beta_{ow}(1 - \rho_{ro})} \qquad (4.13)$$

where $H_o = z_{ao} - z_{ow}$ is the well hydrocarbon thickness. Equation (4.13) indicates that porous medium S-h properties are not needed to compute D_o. Only fluid

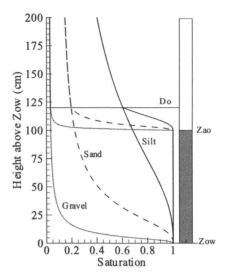

Figure 4.3. LNAPL distribution in three soils for $H_o = 100$ cm. Solid lines denote the water saturation and dashed lines the total liquid saturation. The area between the total liquid and water saturation curve for each soil is an indication of the amount of LNAPL in the profile. The elevation where both saturation curves intersect denotes the upper limit of the soil hydrocarbon thickness, z_{ou}.

density, interfacial tension, and monitoring well hydrocarbon thickness are needed to compute D_o. The independence of D_o from soil properties, when van Genuchten retention parameters are assumed, is shown in Figure 4.3, for three different soils: a sand, silt, and a gravel, for a H_o of 100 cm. Retention parameters associated with these soils as well as fluid properties are shown in Table 4.1. It is obvious that D_o is the same for all three soil types.

Substituting the fluid properties listed in Table 4.1 into Equation (4.13) for a H_o of 100 cm, yields a D_o value of 120 cm for all soils. It is apparent in Figure 4.3 that D_o provides no direct information concerning LNAPL volumes in the subsurface. For instance, the figure shows that although D_o is the same for each soil and independent of van Genuchten retention parameter values, the actual volumes of LNAPL in the three profiles differ considerably.

For Brooks and Corey relations, Equation (4.13) needs to be modified to recognize the LNAPL entry pressure of the porous material. This lower limit can be obtained using the Brooks and Corey function (Equation (2.6)), recognizing that at $z = z_{ol}$, the lower limit, $h_{ow} = h_d/\beta_{ow}$. Thus, the equation for z_{ol} yields:

$$z_{ol} = z_{ow} + \frac{h_d}{\beta_{ow}(1 - \rho_{ro})} \qquad (4.14)$$

TABLE 4.1. Retention parameter values and fluid properties used for soil hydrocarbon thickness and LNAPL specific volume computations.

Porous Media Parameters	Van Genuchten			Brooks-Corey		
	α	n	S_{wir}	h_d	λ	S_{wir}
Gravel	0.352	2.32	0.02	3.5	1.235	0.02
Silt	0.025	2.30	0.13	25.2	0.917	0.13
Sand	0.185	1.65	0.10	6.55	0.535	0.10

LNAPL Properties	
density ratio	0.73
oil-water scaling factor, β_{ow}	1.45
air-oil scaling factor, β_{ao}	3.2

Equation (4.13) for Brooks and Corey S-h relations becomes,

$$D_o = \frac{\rho_{ro}\beta_{ao}H_o}{\rho_{ro}\beta_{ao} - \beta_{ow}(1-\rho_{ro})} - \frac{h_d}{\beta_{ow}(1-\rho_{ro})} \qquad (4.15)$$

The z_{ol} for Brooks and Corey relations (Equation (4.14)) is also sometimes referred to as z_{fow}, which might be interpreted as the upper boundary of the water-LNAPL capillary fringe. The total liquid saturated zone or upper boundary of the LNAPL-air capillary fringe, z_{fao}, is defined as

$$z_{fao} = z_{ao} + \frac{h_d}{\rho_{ro}\beta_{ao}} \qquad (4.16)$$

Equation (4.16) can be obtained from Equation (4.8) using $h_{ao} = h_d/\beta_{ao}$ and solving for z. Figure 4.4 shows fluid saturations as a function of elevation using Brooks and Corey retention relations for the silt in Table 4.1 for the case where $H_o = 100$ cm. While the D_o for the Van Genuchten relation is 120 cm (see also Figure 4.3), the D_o for the Brooks and Corey relation is only 55 cm because of the subtraction of z_{ol} of 65 cm from 120 cm.

The LNAPL volume in the soil per unit area in the horizontal plane (LNAPL specific volume) is given by

$$V_o = \int_{z_{ol}}^{\tau} \phi S_o(z)dz \qquad (4.17)$$

where ϕ is the porosity of the soil, and τ is the minimum of the elevation of the soil surface, z_u, and z_{ou}. Note that V_o represents an effective height of the LNAPL

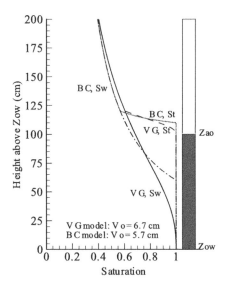

Figure 4.4. Water (solid lines) and total liquid saturations (dashed line) for a silt soil (see Table 4.1 for properties) for the van Genuchten and Brooks and Corey S-h models. The LNAPL well thickness is 100 cm.

zone. Using Equations (4.10), (4.11), and (4.12) to evaluate Equation (4.17) yields an integral expression for the van Genuchten model that needs to be solved using a numerical scheme such as a quadrature algorithms. For the Brooks and Corey model, it is possible to obtain analytical solutions for Equation (4.17). But, first it is recast into the following forms depending on the height z_{fao}:

$$V_o = \phi \int_{z_{ol}}^{z_{fao}} [1 - S_w(z)]\,dz + \phi \int_{z_{fao}}^{z_u} [S_t(z) - S_w(z)]\,dz \quad \text{for } z_{fao} < z_u \qquad (4.18)$$

$$V_o = \phi \int_{z_{ol}}^{z_u} [1 - S_w(z)]\,dz \quad \text{for } z_{fao} \geq z_u \qquad (4.19)$$

The solutions for Equation (4.18) and (4.19) are, respectively,

$$
\begin{aligned}
V_o &= \frac{\phi(1 - S_{wir})}{1 - \rho_{ro}}(C - D) - \frac{\phi(1 - S_{wir})\beta^\lambda (C^{1-\lambda} - B^{1-\lambda})}{(1 - \rho_{ro})(1 - \lambda)} \\
&\quad + \frac{\phi S_{wir}}{\rho_{ro}}(D - E) + \frac{\phi(1 - S_{wir})E^\lambda (D^{1-\lambda} - E^{1-\lambda})}{\rho_{ro}(1 - \lambda)} \\
&\quad + \frac{\phi S_{wir}}{1 - \rho_{ro}}(A - C) - \frac{\phi(1 - S_{wir})B^\lambda (A^{1-\lambda} - C^{1-\lambda})}{(1 - \rho_{ro})(1 - \lambda)}
\end{aligned}
\qquad (4.20)
$$

$$V_o = \frac{\phi(1-S_{wir})}{1-\rho_{ro}}(A-B) - \frac{\phi(1-S_{wir})B^\lambda(A^{1-\lambda}-B^{1-\lambda})}{(1-\rho_{ro})(1-\lambda)} \qquad (4.21)$$

where

$A = (1-\rho_{ro})(z_u - z_{ow})$

$B = h_d/\beta_{ow}$

$C = (1-\rho_{ro})(z_{ao} - z_{ow} + h_d)/(\beta_{ow}\rho_{ro})$

$D = \rho_{ro}(z_u - z_{ao})$

$E = h_d/\beta_{ao}$

The solutions are a function of the *S-h* parameters, H_o, and fluid properties. More details about the derivation of Equations (4.20) and (4.21) can be found in Lenhard and Parker [1990].

The relation between V_o and the LNAPL thickness in observation well, H_o, for a typical silt soil is shown in Figure 4.5. In this figure, the ratio $R = V_o/H_o$ is also depicted. The ratio R is referred to as the LNAPL reduction factor and permits conversion from observation well LNAPL thickness to LNAPL specific volume. The figure shows that the V_o values for both models agree well for larger H_o values. For smaller H_o values, the van Genuchten V_o is considerably larger than the Brooks and Corey V_o. The difference is directly related to distinct

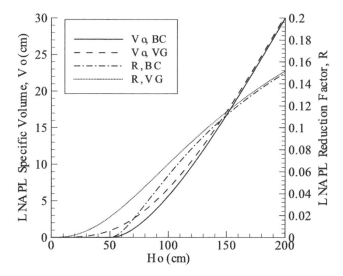

Figure 4.5. Relationships between LNAPL thickness in observation well, H_o, and the LNAPL specific volume, V_o, and the total LNAPL reduction factor, R for van Genuchten and Brooks and Corey retention relations.

non-wetting fluid entry heads that are not included in the van Genuchten model. For the Brooks and Corey model to predict a $V_o > 0$, the following must be true

$$H_o > \frac{h_d \left[\beta_{ao} \rho_{ro} - \beta_{ow}(1 - \rho_{ro}) \right]}{\beta_{ow}(1 - \rho_{ro}) \beta_{ao} \rho_{ro}} \tag{4.22}$$

For very large H_o values, the ratio R will approach $\phi(1 - S_{wir})$.

Charbeneau et al. [1999] argued that for most soils the relationship between V_o and H_o, as shown in Figure 4.5, is approximately linear, except for small H_o values. Instead of the more complex relations (4.20) and (4.21), they proposed the simple relationship

$$V_o = \beta (H_o - \alpha) \tag{4.23}$$

The parameter β in (4.23) represents a capacitance factor for the LNAPL layer, while the parameter α is the intercept of the monitoring well LNAPL thickness axis and can be viewed as an extrapolated estimate of the LNAPL entry head. Based on literature data for retention parameter values, Charbeneau et al. [1999] developed an extensive table with α and β values for a number of general soil types.

The relations presented so far are applicable to non-hysteretic fluid displacement processes. Hysteresis as a result of non-wetting fluid entrapment (e.g., LNAPL entrapment by water) and pore geometry effects are not included. Entrapment of LNAPL occurs when the water table rises and the capillary head decreases. During these rises, the amount of free LNAPL decreases as a result of LNAPL entrapment and subsequently the LNAPL thickness in the wells drops. This is a well-documented observation in field studies. The water table rise might be such that all the free LNAPL becomes entrapped. When that happens, the monitoring well contains no observable LNAPL while considerable amounts of organic liquid are present in the soil.

The effects of pore geometry hysteresis without entrapment are illustrated in Figure 4.6 for a water drainage case (Figure 4.6a) and a water imbibition case (Figure 4.6b) for the same porous medium. To estimate the water imbibition conditions shown in Figure 4.6b, the van Genuchten α value was doubled while the Brooks and Corey h_d was halved. For both cases, the same well observation $H_o = 100$ cm is assumed. For the water imbibition case (e.g., as a result of a water table rise), the computed LNAPL specific volumes are more than twice as large as the V_o value for the water drainage case. These results show the importance of knowing the saturation history at a particular site when converting well observations to LNAPL volumes.

For field situations where the total LNAPL volume in the subsurface is approximately constant, water imbibition (water table rise) results in a reduction of the LNAPL well thickness as a result of both pore geometry hysteresis and LNAPL entrapment. During water table rise, the use of non-hysteretic S-h relations leads to an underestimation of the LNAPL volumes in the subsurface because the effects

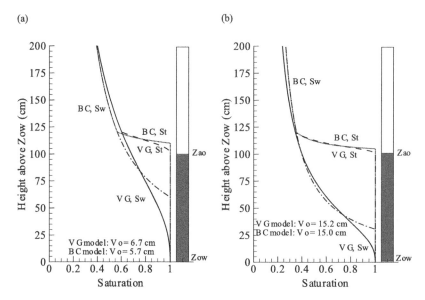

Figure 4.6. LNAPL distribution and volumes for (a) water drainage and (b) water imbibition in a silt for H_o = 100 cm.

of fluid entrapment and pore geometry hysteresis are not accounted for in the retention parameter values. A general overview of hysteretic effects on LNAPL thickness in observation wells is presented by Marinelli and Durnford [1996].

4.1.2   Correlation of LNAPL Well Measurements to LNAPL Volume

Summary: Standard interpolation techniques such as inverse interpolation and kriging can be used to obtain two-dimensional representations of LNAPL specific volumes. Total volumes are subsequently estimated by integrating the specific volumes over the spill area. Example interpolations are completed for two LNAPL sites, indicating that the integration results are strong functions of data density, interpolation techniques, fluid-saturation history at a site, and computational domain boundaries. Interpolation and integration schemes for specific LNAPL volumes have to be used with great caution.

Interpolating LNAPL specific volumes near observation wells may result in useful two-dimensional aerial representations. In addition, the interpolated values can be integrated to yield a total site LNAPL volume. However, improper interpolation, especially using scarce data, may produce unreliable results [Charbeneau, 2000]. The field data may be interpolated and integrated over a specified domain using various software packages. Most interpolating software packages offer linear, inverse distance, and kriging interpolation options. For illustrative purposes, example interpolations in this section are obtained using the inverse distance and kriging options only.

The inverse distance algorithm is relatively simple. The value of a variable at a "destination" data point is calculated as a function of selected "source" (measured) data points. The value at each destination data point (subscript d) is weighted by the inverse of the distance between the source data point (subscript s) and the destination data point raised to a certain power. For LNAPL specific volumes, the relationship is as follows:

$$V_{o,d} = \frac{\sum w_s V_{o,s}}{\sum w_s} \qquad (4.24)$$

where $V_{o,d}$ and $V_{o,s}$ are LNAPL specific volumes at the destination and source point, respectively, the summations are over the source points, and w_s is the weighting function defined as

$$w_s = D^{-E} \qquad (4.25)$$

where D is the distance between the source point and the destination point or a specified minimum distance and E is an exponent with values usually ranging between 2 and 5.

The kriging method is more complex than the inverse-distance method. Generally, kriging produces results superior to the inverse-distance method but requires more computational effort. A detailed description of the method can be found in many textbooks such as Gelhar [1993]. Most interpolating software packages offer users a number of options such as the density of the computational grid, the number of source points used for interpolation at a destination point (N), the type of data drift, and the value of the weighting exponent (for inverse-distance interpolation), and the semi-variance at each source data point (kriging).

In this section, two examples of site LNAPL volumes are presented. The first example is based on field data described by Cooper et al. [1995] and Kaluarachchi and Elliott [1995]. The second example, created by A.S. Mayer for teaching purposes at Michigan Technological University, is related to a hypothetical LNAPL spill site. The field site described by Cooper et al. [1995] contains an estimated spill volume of 215 m^3. After the spill, 16 observation wells were placed (Figure 4.7). The observed well LNAPL thickness values were rather constant over time, ranging from 0 to 0.82 m. The site is considered to be fairly homogeneous allowing the use of one set of retention parameter values. Measured and assumed retention parameter values, porosity, and fluid properties are listed in Table 4.2. Using the values from this table, observed LNAPL well thicknesses were converted to LNAPL specific volumes in the subsurface adjacent to an observation well using a numerical evaluation of Equation (4.17) in Section 4.1.1.

The computed LNAPL specific volumes were then used in interpolation algorithms to simulate aerial distributions and integrated volumes. The purpose of

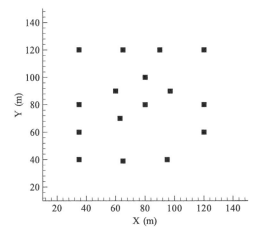

Figure 4.7. Location of observation wells at Cooper et al. [1995] field site.

these simulations is to demonstrate the sensitivity of LNAPL volume estimates to the interpolation technique. Table 4.3 shows the methods, parameters and options used for a number of aerial interpolations of the Cooper et al. [1995] field data, along with the corresponding integrated site LNAPL volume.

According to expectation, a higher grid density results in a smoother variable distribution for both techniques. Examples of a coarse and fine grid for inverse distance method are shown in Figure 4.8 and Figure 4.9, respectively. An increase in the considered source points from 8 to 16 (Interpolation 3) only slightly altered the distribution. Changing the values of the exponents results in an increased relative importance of each source point for lower values (Interpolation 4) and more data smoothing for large value (Interpolation 5). A high value of the exponent yielded negative specific volumes at the edges of the domain, resulting in a lower integrated LNAPL volume.

TABLE 4.2.   Soil and fluid parameter values for the Cooper et al. [1995] field site.

Parameter	Value
porosity, ϕ	0.41
van Genuchten α (m^{-1})	7.5
van Genuchten n	2.5
irreducible saturation S_m	0.2
LNAPL relative density, ρ_{ro}	0.8
air-oil scaling factor, β_{ao}	3.2
oil-water scaling factor, β_{ow}	1.45

TABLE 4.3.   Interpolation information and computed integrated volume.

Interpolation	Method	Grid	Imposed Change with respect to Standard Interpolation*	Integrated Volume (m³)
1	Inverse Distance	15 × 15	None	194
2	Inverse Distance	150 × 150	None	206
3	Inverse Distance	150 × 150	$N = 16$; nearest neighbor	215
4	Inverse Distance	150 × 150	$E = 2$	278
5	Inverse Distance	150 × 50	$E = 5$	179
6	Kriging	15 × 15	None	254
7	Kriging	15 × 15	$N = 8$	134
8	Kriging	150 × 150	None	254
9	Kriging	150 × 150	nearest neighbor	254
10	Kriging	150 × 150	quadratic drift	191
11	Kriging	150 × 150	imbibition parameters	557

*Standard interpolation for the inverse-distance method involves $N = 8$, linear drift, octant coordinate system, and weighting exponent of 3.5. Standard interpolation for kriging involves $N = 16$, linear drift, octant coordinate system. Both standard method use main drainage van Genuchten retention relations, except Interpolation 11. Interpolation 11 was completed using a van Genuchten $\alpha = 15$ m^{-1} to simulate an imbibition saturation path.

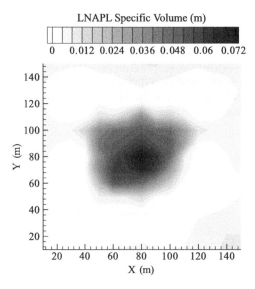

Figure 4.8. Interpolated LNAPL specific volumes using the inverse distance method with 15 × 15 nodes. Used options were 8 considered measured points in octants, linear drift, and a weighting exponent of 3.5.

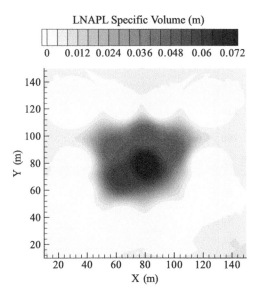

Figure 4.9. Interpolated LNAPL specific volumes using the inverse-distance method with standard settings and a 150×150 node grid.

The kriging method has been used in Interpolations 6-11. A standard interpolation for a 15×15 node grid is shown in Figure 4.10. Note that the standard kriging interpolation (Interpolation 6) yields $60 \, \text{m}^3$ more than the standard inverse distance interpolation (Interpolation 1). Interestingly, Interpolations 6, 8 and 9 yield nearly the same integrated volume of LNAPL. These results suggest that refining the destination grid from 15×15 to 150×150 nodes does not considerably affect the areal LNAPL distribution. Reducing the amount of source points, taken into consideration to compute destination point values, from 16 to 8 (Interpolation 7; Figure 4.11) produced a large zone with negative values. As a result, the integrated volume dropped to $134 \, \text{m}^3$. It is important to recognize that the occurrence of negative values may not be apparent when the results are plotted using contour levels with an expected range from 0 to 0.072 m. It is obvious that this interpolation is not appropriate for the data set. Negative interpolated volumes are also computed when the quadratic drift option is used (Interpolation 10).

Interpolations 1-10 are based on LNAPL specific volumes derived from LNAPL well thicknesses using main drainage non-hysteretic retention parameter values. If, however, main drainage or non-hysteretic conditions are not appropriate for a certain site, the computed volumes might be in error. As an example, an interpolation has been completed using assumed main imbibition values for the site soil. When doing that, the LNAPL specific volumes are more than two

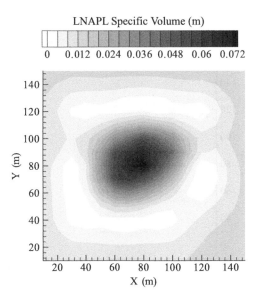

Figure 4.10. Interpolated LNAPL specific volumes using kriging with a 15×15 node grid. Used options were 16 considered measured points in octants and linear drift.

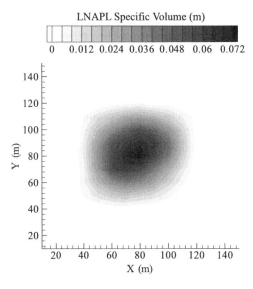

Figure 4.11. Interpolated LNAPL specific volumes using kriging with a 15×15 node grid. Used options were 8 considered measured points in octants and linear drift. Specific volume contour labels range from –0.08 to 0.072.

TABLE 4.4. Well coordinates and LNAPL well thickness data for the Mayer example.

x (m)	y (m)	LNAPL well thickness (m)
4.77	0.66	0.00
9.52	0.55	0.00
16.23	0.68	0.01
5.99	3.46	1.01
15.98	4.17	0.33
3.45	6.20	1.22
6.71	6.00	1.53
5.40	8.08	1.83
8.82	8.17	1.32
7.92	9.40	1.34
15.45	9.31	0.41
4.69	12.19	1.28
3.40	12.70	1.08
8.75	12.85	1.13
5.39	13.41	0.93
15.27	13.29	0.71
1.49	9.51	0.93

times larger near each observation well than the volumes based on main drainage retention parameter values. As a result, the integrated volume for the whole site increases to over 500 m^3.

The hypothetical data set developed by Mayer is given in Table 4.4 and Table 4.5. In this case, the Brooks and Corey relations are used to convert LNAPL well thickness data to specific LNAPL volumes using Equations (4.20) and (4.21). Kriging interpolations using a 150×150 node grid for the LNAPL well thickness, soil hydrocarbon thickness, and specific LNAPL volumes are shown in Figure 4.12, Figure 4.13, and Figure 4.14, respectively, for an area of 17×14 m.

TABLE 4.5. Soil and fluid properties for the Mayer example.

Parameter	Value
porosity, ϕ	0.43
Brooks and Corey h_d (m)	0.077
Brooks and Corey λ	0.924
irreducible saturation S_{wir}	0.0
LNAPL relative density, ρ_{ro}	0.73
air-oil scaling factor, β_{ao}	1.46
oil-water scaling factor, β_{ow}	3.04

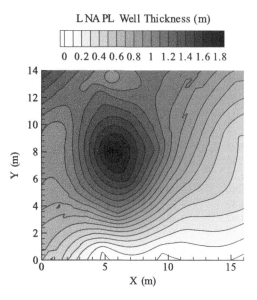

Figure 4.12. Interpolated LNAPL well thickness (m) using standard kriging method for Mayer's data set.

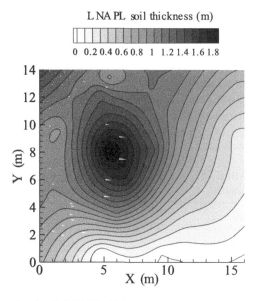

Figure 4.13. Interpolated soil LNAPL thickness (m) using standard kriging method for Mayer's data set.

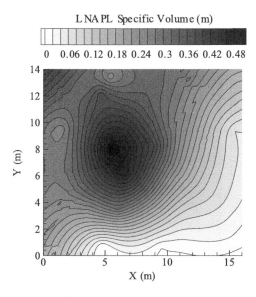

Figure 4.14. Interpolated specific LNAPL volume (m) using standard kriging method for Mayer's data set.

Figure 4.13 is created with soil hydrocarbon thickness data computed with Equation (4.15), while Figure 4.14 is made with the specific LNAPL volumes calculated using Equations (4.20) and (4.21). The aerial integration of the specific LNAPL thickness yields a total volume of 45 m^3, which represents an estimate of the free LNAPL volume present in the domain. A similar integration of the soil hydrocarbon thickness yields a value of 160 m^3. This quantity represents the volume of soil contaminated with continuous LNAPL and not the LNAPL volume. The large difference between the two values indicates that it is important for these two variables to be used in the proper physical context.

Another issue complicating aerial interpolations and integrations is the uncertainly in defining the boundaries where the well LNAPL thickness is zero. In this particular example, the computational domain used in Figure 4.12, Figure 4.13, and Figure 4.14 was 17×14 m. It is certainly possible that LNAPL might be present outside the chosen domain. An example of what might happen to the interpolated specific volumes if another aerial domain is selected, is shown in Figure 4.15 for a 30×30 m domain. The scarcity of data beyond $x = 16$ m and $y = 13$ m results in some awkward iso-saturation patterns in that zone. The computed free volume of LNAPL for this figure is 63 m^3, indicating that the computational domain choice can have a large impact on the integrated volumes.

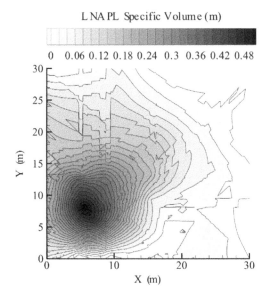

Figure 4.15. Interpolated specific LNAPL volume (m) using standard kriging method for Mayer's data set and a 30 × 30 m domain.

4.2   OBSERVATIONS OF DISSOLVED NAPL COMPONENTS

4.2.1   NAPL Components Present in Groundwater at Lower than Solubility Concentrations and Fluctuations in Concentrations

Summary: In samples of groundwater taken from monitoring wells, concentrations of dissolved NAPL components are usually found at significantly less than solubility concentrations. Potential explanations for these low concentrations include dilution effects, mass transfer limitations, and multicomponent (Raoult's law) effects. Observed fluctuations in dissolved NAPL component concentrations can be explained by multicomponent effects and variations in infiltration rates through NAPLs trapped in the vadose zone.

In this section, we discuss observations made from groundwater sampling that are related to the presence of NAPLs at a contaminated site. When we consider the dissolution of the NAPL into the aqueous phase, or groundwater, we are concerned with quantifying the NAPL as a source of contamination of the groundwater. Figure 4.16 illustrates DNAPL and LNAPL releases and possible scenarios where NAPL dissolution into flowing groundwater can occur. We see that dissolution occurs in the vadose zone, where infiltrating water passes through zones contaminated with residual NAPL, and the saturated zone. For the case of LNAPLs found on top of the capillary fringe/water table, NAPL dissolution can

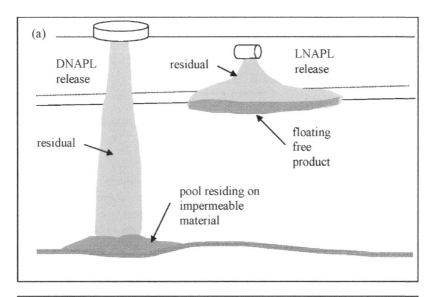

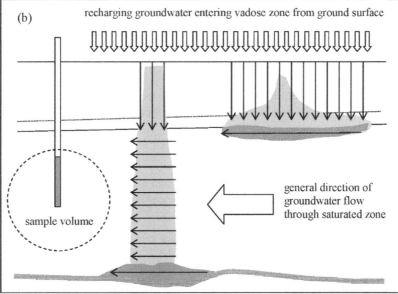

Figure 4.16. NAPL dissolution scenarios: (a) DNAPL and LNAPL contamination and (b) NAPL contact with flowing groundwater.

occur at the interface between the floating LNAPL and the underlying, flowing groundwater. For the case of DNAPLs in the saturated zone, groundwater flows through residual DNAPL and around and through DNAPL pools that have collected on strata that are impermeable to the DNAPL.

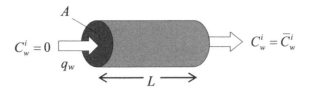

Figure 4.17. Simple conceptual model of single component NAPL dissolution, under one-dimensional flow conditions.

We begin with a simple conceptual model of NAPL dissolution, where we assume that these dissolution scenarios can be modeled as systems where the groundwater flow through the NAPL zone is steady and one-dimensional (see Figure 4.17). We further assume that the NAPL and flowing groundwater are in complete physical contact with each other and are at instantaneous chemical equilibrium with each other. This assumption, known as the local equilibrium assumption (or LEA) leads to immediate determination of the groundwater concentration as

$$C_w^i = \chi_o^i \bar{C}_w^i \qquad (4.26)$$

where $\bar{C}_w^i$ is the aqueous solubility of a pure compound i and χ_o^i is the mole fraction i in the NAPL. In the case of a single-component NAPL ($\chi_o^i = 1$), Equation (4.26) reduces to $C_w^i = \bar{C}_w^i$.

Given the specific discharge of groundwater through the NAPL as q_w, and assuming a single-component NAPL, the mass flux of dissolved NAPL leaving the NAPL zone is

$$J_w^i = q_w \bar{C}_w^i \qquad (4.27)$$

If we know the initial saturation of NAPL present (S_o), the porosity (ϕ), and the length (L) of the NAPL-contaminated zone, we can determine the time necessary for complete dissolution of the NAPL:

$$t^* = \frac{S_o \phi \rho_o L}{J_w^i} = \frac{S_o \phi \rho_o L}{q_w \bar{C}_w^i} \qquad (4.28)$$

Further, we can estimate the minimum volume of groundwater contaminated by the dissolving NAPL, given the cross sectional area normal to the direction of flow (A):

$$(V_w)_{total} = \frac{S_o \phi \rho_o L A}{\bar{C}_w^i} \qquad (4.29)$$

This volume represents only the volume passing through NAPL zone and attaining concentration of $\bar{C}_w^i$. If we take dispersion into account, the actual volume of contaminated groundwater may be much larger, but at a lower concentration.

Example: Given the following information

$\rho_o = 1.63 \ g/cm^3$ $\bar{C}_w^i = 150 \ mg/L$

$S_o = 0.20$ $L = 10 \ m$ $A = 1 \ m^2$

$\phi = 0.3$ $q_w = 0.1 \ m/day$

calculate the time required for complete dissolution and volume of contaminated groundwater.

Solution:

$$t^* = \frac{S_o \phi \rho_o L}{q_w \bar{C}_w^i} = \frac{(0.20)(0.3)(1.63 g/cm^3)(10m)}{(0.1 m/day)(150 mg/L)}$$

$$\times \left(\frac{1000 \, mg}{g} \right) \left(\frac{1000 \, cm^3}{L} \right) \left(\frac{yr}{365 \, day} \right)$$

$$= 178 \ yr$$

$$(V_w)_{total} = \frac{S_o \phi \rho_o LA}{\bar{C}_w^i} = \frac{(0.20)(0.3)(1.63 g/cm^3)(10 m)(1 m^2)}{(150 mg/L)}$$

$$\times \left(\frac{1000 \, mg}{g} \right) \left(\frac{1000 \, cm^3}{L} \right) \left(\frac{1000 L}{m^3} \right)$$

$$= 6.52 \times 10^6 \ L$$

We note from Equations (4.28) and (4.29) that both the time necessary for complete dissolution and the volume of contaminated groundwater are inversely related to the solubility concentration, $\bar{C}_w^i$. As we discuss in the following, the concentration of dissolved NAPL leaving the NAPL-contaminated region is usually orders of magnitude lower than the solubility level. This implies that the time for complete dissolution and volume of contaminated groundwater will be orders of magnitude greater than that predicted using the solubility concentration.

The simple model we have adopted up to this point implies that we can expect groundwater concentrations observed at NAPL-contaminated sites to be near solubility levels and that the rate of NAPL dissolution is constant. In the following, we discuss several phenomena- *mass transfer limitations, heterogeneity, multicomponent (Raoult's law) effects, dilution effects,* and *temporal variations in infiltration rates*- which can render the simple model inappropriate. Of course, in addition to these phenomena, spatial and temporal variations in dissolved NAPL concentrations also can arise because of variability in biogeochemical processes such as biodegradation and sorption.

Mass transfer limitations: Previously, we invoked the local equilibrium assumption (LEA), which implies that groundwater concentrations immediately reach the solubility limit under any circumstance where flowing groundwater is in contact with NAPL. However, this is rarely the case, even under homogeneous conditions. Laboratory experiments involving single-component NAPLs have demonstrated that solubility conditions are not maintained in groundwater immediately downstream of a residual or pool zone as it dissolves. Rather, long tailing occurs following a relatively short period of time during which concentrations are near solubility. Figure 4.18, for example, illustrates data collected by Mayer et al. [1999] on the dissolution of NAPL residual in one-dimensional sand columns. After NAPL (trichloroethylene, or TCE) was introduced to the column as residual, the column was flushed with clean water and the concentration of TCE in the effluent was measured.

The horizontal axis in Figure 4.18 represents the amount of water flushed through the column in pore volumes. The figure clearly shows that trichloroethylene saturation is maintained for one hundred or so pore volumes, after which a long tail is observed. The tailing is in part the result of a reduced NAPL-water interfacial area as the NAPL dissolves and the diffusion-controlled dissolution of NAPL situated in pores and pore throats inaccessible to the flowing groundwater.

Figure 4.19 shows a pore-scale representation of the dissolution process, where the pore size and distribution are arranged so that the majority of the groundwater flow occurs in the lower pore. Initially (Figure 4.19(a)), the portion

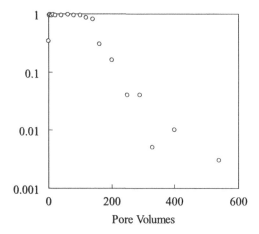

Figure 4.18. Effluent groundwater concentration, normalized by solubility, versus pore volumes flushed, where an experimental column containing trichloroethylene was flushed with clean water at a pore velocity of 1.1 m/day.

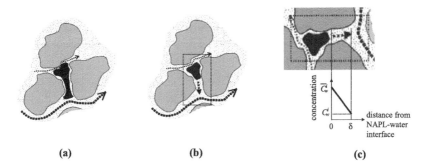

Figure 4.19. Schematic illustration of the dissolution of a NAPL blob at the pore scale at (a) early time, (b) later time, and (c) conceptual representation of mass transfer model.

of the NAPL blob adjacent to the lower pore will dissolve preferentially, due to the greater flux of water through that pore. Since there is direct contact between the blob and the water flowing through the lower pore, the dissolution is relatively fast. However, at later times (Figure 4.19(b)), the blob has dissolved such that there is no longer direct contact between the flowing water and the blob. The transport of dissolved NAPL to the water flowing through lower pore occurs by only diffusion from the blob through the pore throat and into the lower pore, resulting in a significantly lower dissolution rate. This phenomenon gives rise to the tailing or mass-transfer limitation phenomenon observed in the data in Figure 4.18.

A linear driving force model is often used to represent mass-transfer limited dissolution. Using this conceptual model, the dissolution rate J_w^i is described as diffusion across a stagnant water film of width δ (Fick's first law), as in

$$J_w^i a = k_{mt}^i a(\bar{C}_w^i - C_w^i), \quad k_{mt}^i = D_w^i / \delta \qquad (4.30)$$

where k_{mt}^i is the mass transfer coefficient, a is the specific interfacial area between the water and NAPL per unit volume of porous medium (units of L^2/L^3) and D_w^i is the free liquid diffusivity in water. The mathematical model in Equation (4.30) can be related to the conceptual model illustrated in Figure 4.19, where $\bar{C}_w^i$ is the concentration at the interface between the blob and the water in the pore throat, C_w^i is the bulk concentration in the water flowing through the lower pore, and δ is the distance between the NAPL blob-water interface and the lower pore. The driving force for the dissolution is the difference $\bar{C}_w^i - C_w^i$, proportioned by the product of the mass transfer coefficient, k_{mt}^i and interfacial area, a, at the NAPL-blob interface.

The mass transfer coefficient usually is estimated from one-dimensional column test experiments with a setup similar to that shown in Figure 4.17. The length scale of these experiments is on the order of a few centimeters. Since the interfacial area cannot be measured independently, a lumped mass transfer rate coefficient, $k_{mt}^{i*} = k_{mt}^i a$, is estimated. The measured mass transfer rate coefficient is usually related to system properties using empirical models, such as

$$Sh = \frac{k_{mt}^{i*} d_p^2}{D_w^i} = \beta_0 \theta_o^{\beta_1} Re^{\beta_2} \qquad (4.31)$$

where Sh is the dimensionless Sherwood number, d_p is the characteristic particle size, $Re = v_w \rho_w d_p / \mu_w$ is the Reynolds number, and β_i are fitting coefficients. Additional parameters have been used in these empirical models, such as other characteristics of the grain size distribution [Powers et al., 1994] and distance into the mass transfer zone [Imhoff et al., 1994]. A model such as that in Equation (4.31) allows estimation of k_{mt}^{i*} as a function of fixed properties, such as porous media characteristics, and properties that will vary during dissolution, such as the NAPL volumetric fraction and pore velocity. We note that these empirical models were developed from centimeter-scale column experiments; however, k_{mt}^{i*} is undoubtedly dependent on spatial scale. At this time, very little work has been done on upscaling functional relationships and measurements at the pore or column scale to the practical modeling scale.

The mass transfer-limited dissolution model can be incorporated into the water mass balance (advective-dispersive) equation, as follows [Miller et al., 1990]:

$$\frac{\partial(\theta_w C_w^i)}{\partial t} = -q_w \frac{\partial C_w^i}{\partial x} + \theta_w D_w^i \frac{\partial^2 C}{\partial x^2} + \theta_w k_{mt}^{i*}(\bar{C}_w^i - C_w^i) \qquad (4.32)$$

For the case where the mass of NAPL is changing slowly, we do not need a mass balance equation for the NAPL and the conditions are close to steady state. Further, if dispersion can be neglected, a simple analytical solution to Equation (4.32) under steady state conditions ($\partial C/\partial t = 0$) can be derived [Miller et al., 1990]:

$$\frac{C_w^i}{\bar{C}_w^i} = 1 - \exp\left(\frac{-k_{mt}^{i*}}{v_w^i} x\right) \qquad (4.33)$$

using a boundary condition of $C_w^i = 0$ at $x = 0$. Note that the dimensionless ratio $k_{imt}^* x / v_w$ is also known as the Damkohler number, Da.

Example: Equation (4.33) can be used to determine the length of NAPL-contaminated region, or mass transfer zone, required to produce equilibrium or near equilibrium conditions; for example, we could choose a ratio $C_w^i / \bar{C}_w^i = 0.99$ to represent equilibrium conditions. In Miller et al. [1990] a mass transfer rate coefficient of $k_{mt}^{i*} = 10^{-5}$ m/s was measured with a toluene volumetric *fraction of $\theta_o = 0.1$ and a pore velocity of $v_w = 1.2 \times 10^{-6}$ m/s. For this case, the length of the mass transfer zone, x_{mt}, is*

$$x_{mt} = \frac{v_w}{k_{mt}^{i*}} \ln\left(1 - \frac{C_w^i}{\bar{C}_i^w}\right) = \frac{1.2 x 10^{-6}}{10^{-5}} \ln(1 - 0.99) = 0.53 \ m$$

This length is relatively short compared to the mass transfer length for the situation where a significant amount of NAPL has dissolved. For a lower toluene volumetric fraction of $\theta_o = 0.001$, $k_{mt}^* = 2.5 \times 10^{-7}$ m/s was measured by Miller et al. [1990]. For this case, the mass transfer zone length is $x_{mt} = 21$ m. Furthermore, if we increase the pore velocity to a value that could occur under pumping conditions, say $v_w = 1.2 \times 10^{-4}$ m/s, we have a mass transfer zone of 2,100 m. This example demonstrates that mass transfer limitations may be important for low NAPL volumetric fractions or high pore velocities.

The utility of models described in Equations (4.30)–(4.32) is limited by the fact the equations and parameters are based pore-scale conceptual models and centimeter-scale column experiments.

Heterogeneity. At larger scales, groundwater will not always flow through residual and/or pool zones in as ideal a manner as in laboratory experiments, as demonstrated in Figure 4.20. At this scale, a heterogeneous distribution of hydraulic conductivity will cause groundwater to flow preferentially through coarser-grained lenses and laminations, resulting in less than optimal contact with certain residual and pool zones. In other words, in addition to diffusion-limited mass transfer at the grain scale, an analogous diffusion-controlled mass transfer may occur at the macroscopic scale, again producing long tailing.

Figure 4.21 presents the results of a series of simulations conducted by Mayer et al. [1996] of the dissolution of a hypothetical NAPL release into an aquifer. The graph shows the mass of NAPL remaining in the hypothetical aquifer, m_o, normalized by the initial NAPL mass, versus dimensionless time, or pore volumes flushed through the aquifer. Each simulation was conducted with a different level of porous media heterogeneity. The level of heterogeneity is indicated by the variance of the log of the hydraulic conductivity, σ. A variance of $\sigma = 0$ indicates a homogeneous system. The lowest non-homogeneous variance, $\sigma = 0.29$, indicates about a three-order-of-magnitude range of hydraulic conductivity; the highest variance, $\sigma = 2.32$, indicates a seven-order-of-magnitude range. The results in Figure 4.21 indicate that the time required for 99% $(m_o/(m_o)_{initial} = 10^{-2})$ removal of NAPL increases roughly in proportion to the level of heterogeneity.

Equation (4.28) and the previous example demonstrate that the low solubility of NAPL compounds results in long life spans for NAPL releases, even when equilibrium conditions are assumed. Mass transfer limitations have the effect of

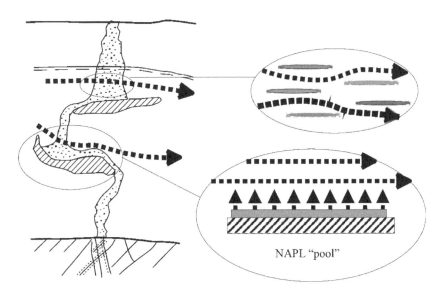

Figure 4.20. Illustration of the effects of field-scale heterogeneity on water-NAPL contact.

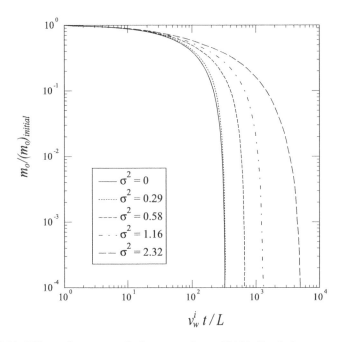

Figure 4.21. Effects of porous media heterogeneity on NAPL dissolution.

significantly increasing the life span of NAPL releases, relative to equilibrium conditions, as indicated by the tailing in Figure 4.18. Furthermore, Figure 4.21 shows that the presence of heterogeneity will increase the time required to dissolve a NAPL release, by orders of magnitude. Combined together, solubility limits, mass transfer limitations, and heterogeneity can produce conditions where NAPLs will take decades to centuries to dissolve.

The schematic illustration in Figure 4.22 describes a conceptual model of dissolution from NAPL pools. NAPL saturations in pools can be significantly higher than residual levels, giving rise to low water relative permeabilities within the pool, further resulting in low water fluxes through the pool. In this case, the dominant removal NAPL removal mechanism may be dissolution from the top of the pool. Figure 4.22 illustrates three steps in a simple conceptual model of NAPL pool dissolution: (1) equilibrium between the NAPL in the pool and the groundwater at the pool/water interface, resulting in solubility concentrations in the groundwater at the interface; (2) dispersion of the NAPL dissolved in the groundwater away from the pool interface, in the direction transverse to the pool length; and (3) advection of the dissolved NAPL in the direction of bulk groundwater flow, which may be more or less parallel to the pool length.

The concentration profile along the pool interface given in Figure 4.22 indicates that the dissolved NAPL concentration is zero at the upstream end of the pool and gradually increases along the length of the pool, due to greater contact between the flowing groundwater and the pool. The concentration profile along the direction transverse to the pool interface indicates that the dissolved concentration is equal to the solubility concentration at the interface and decreases away from the interface, due to dilution effects. The vertically-average concentration leaving the pool could be near or at equilibrium concentrations, depending on factors such as the residence time along the pool and the transverse dispersion.

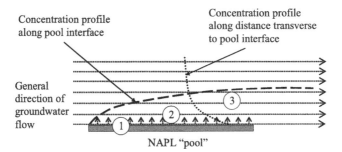

Figure 4.22. Conceptual model of NAPL pool dissolution, including (1) equilibrium between the NAPL and groundwater at the pool/water interface, (2) dispersion of the dissolved NAPL away from the pool interface, and (3) advection of the dissolved NAPL in the direction of bulk groundwater flow.

Hunt et al. [1988] derived a solution for the concentration profile in the direction transverse to a pool at the downstream edge of the pool:

$$\frac{C_w^i}{\overline{C}_w^i} = \text{erfc}\left(\frac{z}{2\sqrt{D_{w,t}^i L \big/ v_x}}\right) \quad (4.34)$$

where erfc is the complementary error function, z is the distance transverse to the pool ($z = 0$ at the pool/groundwater interface) $D_{w,t}^i = \alpha_t v_x + \tau_w D_w^i$ is the dispersion transverse to the length of the pool, α_t is the transverse dispersivity, v_x is the velocity in the direction parallel to the length of the pool, and L is the pool length.

Johnson and Pankow [1992] derived a dissolution rate, averaged over the pool length, as

$$J_w^i = \phi \overline{C}_w^i \left[(4 D_{w,t}^i v_x) \big/ (\pi L) \right]^{1/2} \quad (4.35)$$

If it is assumed that the areal dimensions of the pool do not vary during dissolution (in other words, the pool dissolves only from "top to bottom"), the lifetime of the pool can be estimated as

$$t^* = \frac{DL\phi S_o \rho_o}{J_w^i} \quad (4.36)$$

where D is the depth of the pool.

According to Equations (4.34), (4.35) and (4.36), the dissolved concentration in the groundwater leaving the pool, the NAPL dissolution rate and the pool lifetime depend on parameters such as the solubility concentration at the interface, the transverse dispersion, and the groundwater flow rate, in addition to the configuration of the NAPL pool. Of these parameters, perhaps the most difficult to quantify is the transverse dispersion, which depends on chemical diffusion and transverse mechanical dispersivity. The transverse mechanical dispersivity is especially difficult to ascertain, given that it cannot be measured reliably in situ or determined theoretically.

Example: Use Equations (4.34), (4.35) and (4.36) to determine the sensitivity of dissolved concentrations leaving the NAPL pool and the NAPL pool lifetime to the transverse dispersivity, for a range of pore velocities. Use pool dimensions of $L \times H = 5\ m \times 0.05\ m$; NAPL characteristics of $\overline{C}_w^i = 1,100\ mg/L$, $\rho_o = 1.46\ g/cm^3$, $D_{w,t} = 8 \times 10^{-6}\ cm^2/s$, $S_o = 0.3$; porous medium characteristics of $\phi = 0.3$ and $\tau = 0.7$, and pore velocities of $v_x = 0.1$, 1.0, and 10 m/day. The NAPL characteristics are similar to those found for trichloroethylene.

Solution: Figure 4.23 shows the normalized concentrations at the downstream edge of the pool as a function of distance above the pool for the given parameters and transverse dispersivities of 0.01, 0.1, and 1 cm. These results show that the concentrations leaving the pool are quite sensitive to the transverse dispersivity.

Table 4.6 shows NAPL pool lifetimes for the given parameters and transverse dispersivities of 0.01, 0.1, and 1 cm. These results also show that the pool lifetimes are not only sensitive to the transverse dispersivity but also are to the pore velocity. The sensitivity to pore velocity is not unexpected, since the greater the groundwater flow rate above the pool, the faster the dissolved NAPL is carried away from the pool. The results in Table 4.6 indicate that NAPL pool lifetimes can be very long for low pore velocities and transverse dispersivities.

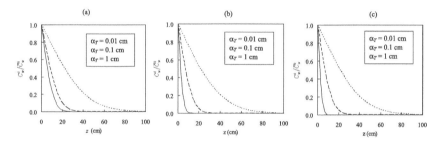

Figure 4.23. Normalized concentrations as a function of transverse dispersivity (numbers indicated in legend) and pore velocity = (a) 0.1 m/day, (b) 1.0 m/day, and (c) 10 m/day. Concentrations calculated from Equation (4.34). Parameters used to calculate concentrations include $L = 5$ m, $D_{w,t} = 8 \times 10^{-6}$ cm^2/s and $\tau = 0.7$.

TABLE 4.6.   NAPL pool lifetime as a function of transverse dispersivity and pore velocity. Lifetime calculated from Equation (4.36). Parameters used to calculate concentrations include pool dimensions of $L \times H = 5$ m $\times 0.05$ m; NAPL characteristics of $\overline{C}_w^i = 1{,}100$ mg/L, $\rho_o = 1.46$ g/cm^3, $D_{w,t} = 8 \times 10^{-6}$ cm^2/s, $S_o = 0.3$; porous medium characteristics of $\phi = 0.3$ and $\tau = 0.7$.

NAPL pool lifetimes (yr)		Transverse dispersivity (cm)		
		0.01	0.1	1
Pore	0.1	149	93.5	35.2
velocity	1.0	29.6	11.1	3.59
(m/day)	10	3.52	1.14	0.36

Multicomponent effects. In addition to hydrodynamic influences such as advection and dispersion, the concentration of a particular contaminant in groundwater downstream of either residual or pooled NAPL will be a function of the component composition of the NAPL. For structurally similar compounds, as described by Equation (2.42), it has been found that the maximum concentration of a given component in groundwater in contact with NAPL will be proportional to the mole fraction of the component in the NAPL [Mackay et al., 1991; Banerjee, 1984). This is a statement of Raoult's Law, which stipulates that the effective solubility of the component in question will equal the product of its single-component solubility and its mole fraction in the NAPL phase.

Consider a two-component NAPL with equal mole fractions (0.5), where the solubilities of the pure-phase compounds are 1,000 mg/l and 10 mg/l, respectively, and the effective solubilities are 500 mg/l and 5 mg/l, respectively. The more soluble compound will potentially partition into ground water 100-fold

TABLE 4.7. Concentrations of hydrocarbon components measured in groundwater mixed with JP-4 jet fuel and gasoline.

Component	Concentrations in Water in Contact with Fuel ("effective solubilities")[b]			
	Pure-Phase Water Solubility[a] (mg/L)	JP-4[c] (mg/L)	Gasoline[d,f] (mg/L)	MCL[e,f] (mg/L)
Benzene	1,750	9.82	58.7	0.005
Toluene	524	8.49	33.4	1.0
Ethylbenzene	187	0.67	4.3	0.7
o-Xylene[g]	167	1.21	6.9	NA
m-Xylene	157	2.01	11.0	NA
p-Xylene	180	0.41	4.4	NA
Xylenes	168	3.63	22.3	10
Trimethylbenzenes[g]	97.7	0.87	1.1[h]	NA
Naphthalene	22	0.39	NA	0.02[i]
Methylnaphthalenes	25.4[j]	0.24	NA	NA

[a]Solubilities at 25°C [Montgomery, 1996].
[b]Fuel to water ratio 1:10.
[c]Smith et al., 1981.
[d]*American Petroleum Institute* [1985].
[e]MCL = maximum contaminant level [EPA, 1996].
[f]NA = not applicable.
[g]Solubility for 1,3,5-trimethylbenzene.
[h]Value for 1,2,4-trimethylbenzene.
[i]Health advisory value for 70-kilogram adult, lifetime exposure.
[j]Value for 2-methylnaphthalene.

more readily than the less soluble compound. Accordingly, less soluble compounds will primarily be associated with the NAPL phase and dissolution and transport in the aqueous phase will be limited relative to more soluble components. Table 4.7 shows the concentrations of various hazardous hydrocarbon components measured in groundwater mixed with JP-4 jet fuel and mixed with gasoline. All of the compounds are found at significantly lower concentrations than the solubility concentration for the pure compound. The data in Table 4.7 also indicate that the compounds with the highest solubilities (benzene and toluene) are found at the highest concentrations. Note that the majority of the jet fuel and gasoline is composed of low molecular weight hydrocarbons with very low solubilities. Multicomponent dissolution also leads to preferential dissolution of the more soluble components out of the NAPL earlier in time, leaving behind the less soluble components to dissolve more slowly. This phenomenon is demonstrated in the following example.

Example: Figure 4.24 shows the results for a simulation of the dissolution of a NAPL representing a five-component, hydrocarbon mixture. In this simulation, we model the dissolution of hydrocarbons in groundwater flowing uniformly through a block of aquifer material uniformly contaminated with the NAPL. The initial composition of the NAPL is given in Table 4.8. The most prevalent component, labeled "Insoluble" in Table 4.8, is a mixture of insoluble compounds. The groundwater flow rate is 100 L/day through a block with an area of 0.1 m². In Figure 4.24 we see that benzene, the component with the highest solubility (1,750 mg/L) and initial mole fraction, is removed quickly from the simulated zone of contamination. Toluene, with the second highest solubility and initial mole fraction, is removed next, and so on. Ethylbenzene and the xylenes, which are present initially at relatively low mole fractions and have somewhat lower solubilities, dissolve from the mixture very slowly, until the benzene and toluene have been removed.

A spreadsheet file (multi.xls) that can be used to make calculations of mass removal from a multi-component NAPL via dissolution is found on the accompanying CD.

Monitoring wells located downstream of multicomponent NAPL sources should show similar behavior: declining concentrations over time of the more soluble components, with perhaps relatively unchanging or slightly increasing concentrations over time for the less soluble components. If multicomponent dissolution is not considered when interpreting contaminant concentrations from such a site, the overall low concentrations of all components and the decline in concentration over time of certain components may be taken as a false sign that NAPL is not present in the subsurface.

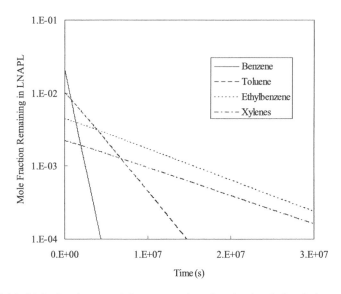

Figure 4.24. Mole fraction remaining versus time for simulated dissolution of a five-component NAPL.

TABLE 4.8.   Initial NAPL Composition Used with Simulation Results in Figure 4.24.

Component	Initial Mass (gm)	Molecular Weight (gm/mol)	Initial Moles	Initial Molar Fraction	Pure-Phase Solubility (mg/L)	Initial Effective Solubiligy (mg/L)
"Insoluble"	930	45	2.07×10^1	9.62×10^{-1}		
Benzene	35	78	4.49×10^{-1}	2.09×10^{-2}	1,750	36.57
Toluene	20	92	2.17×10^{-1}	1.01×10^{-2}	524	5.307
Ethylbenzene	10	106	9.43×10^{-2}	4.39×10^{-3}	187	0.827
Xylenes	5	106	4.72×10^{-2}	2.20×10^{-3}	168	0.37
Total	1,000		2.15×10^1	1.00		

   While multicomponent dissolution can explain some of the temporal variations in contaminant concentration observed at sites, spatial variations occur as well. At sites where the distribution of hydraulic conductivity is highly variable, it follows that residual and pool zones will be distributed in a complex and sparse manner. Because these zones act as sources for dissolved-phase plumes, it follows that such sites will not contain one large, smoothly varying contaminant plume. Instead, the sites will appear to have several isolated plumes, which in two-dimensional cross section do not appear to be related. While these plumes may actually be connected in three dimensions, it is not surprising to find concentration reversals with depth in any given vertical cross section and even monitoring wells exhibiting non-detectable levels of contamination surrounded by wells showing high levels of contamination both above and below.

Dilution effects: Let us consider what is known as the "1% rule-of-thumb" for the confirmation of the presence of DNAPL. DNAPL is infrequently observed directly in monitoring wells; however, the presence of subsurface NAPL is generally suspected if dissolved contaminants are present in concentrations greater than 1% of their aqueous solubility. The 1% rule-of-thumb was developed for the U.S. EPA during the early 90's as a guideline to signal regulatory personnel about the potential for DNAPL at a site without being particularly informed about the physico-chemical or transport principles involved. On this basis, the 1% rule-of-thumb is a conservative measure designed to "raise the red flag" at sites and trigger further investigation.

The 1% rule-of-thumb arose from empirical observations and scientific explanations, primarily having to do with dilution. The dilution effect is explained in Figure 4.25. By the time the dissolved DNAPL plume reaches the well, some degree of mixing will have taken place due to dispersive processes, resulting in a concentration at the sampling point that is a fraction of the solubility concentration. Furthermore, when the groundwater sample is taken from the groundwater well, it is possible that the sample volume will incorporate not only groundwater with $C_w^i \ll \bar{C}_w^i$, but also groundwater from other areas surrounding the well screen and groundwater in the borehole, each of which could contain dissolved NAPL at lower or even zero fractions of solubility. The result is a considerable

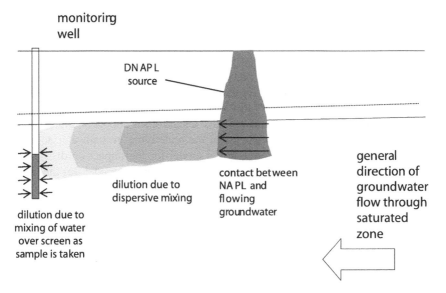

Figure 4.25. Dilution of groundwater sample taken downgradient of the NAPL source area.

dilution between the point where the groundwater flows through the NAPL and the collection of the groundwater sample from the monitoring well.

However, groundwater concentrations less than the 1% solubility screening threshold do not necessarily confirm the absence of NAPL, since monitoring well location, construction, and sampling affect the degree of contaminant dilution. The rule-of-thumb is particularly suspect in some fractured rock settings where a significant fraction of the NAPL mass has diffused into the rock matrix as dissolved mass, and the resulting reverse-diffusion out of the rock matrix and into the fractures has led to groundwater concentrations much lower than 1% of solubility

We can also examine the potential effects of dilution by revisiting the conceptual model of NAPL pool dissolution presented in Figure 4.22. Although we expect concentrations leaving the pool near the pool/groundwater interface to be relatively high, the average, or mixed, concentration leaving the pool will be a product of the concentrations over a vertical mixing length above the pool. The results in Figure 4.23 show that, as expected, the concentrations are close to solubility within the first ones to tens of cm above the pool. The generally low concentrations at short distances above the pool indicate that concentrations downgradient of the pool can be expected to be a small fraction of the solubility, given that mixing within the plume leaving the pool (dispersion) and dilution with uncontaminated groundwater will produce relatively low overall concentrations in the plume leaving the pool.

Temporal variations in infiltration rates. In our simple conceptual model of NAPL dissolution, we assumed that rate of groundwater flow through the region of NAPL contamination was constant. Under these conditions, we should expect to see groundwater concentrations remain steady (except for multicomponent effects). However, it is not unusual to see groundwater concentrations of dissolved NAPL components fluctuate significantly with time, under either natural conditions or conditions where active remediation is occurring. Fluctuating groundwater concentrations are observed in groundwater samples collected periodically from observation wells over months or years.

In cases where the majority of the NAPL is found trapped above the water table, variations in infiltration rates can have a significant impact on the concentration of dissolved NAPL components in the underlying groundwater aquifer. Figure 4.26 illustrates a simple conceptual model for the impact on dissolving NAPL in the vadose zone on the underlying aquifer. In this model, (a) the infiltration rate through the NAPL source can vary with time ($Q = Q_I(t)$), (b) the concentration in the infiltrating water leaving the NAPL is constant with time ($C = C_I$), (c) the flowrates through the aquifer upgradient and downgradient of the NAPL are approximately equal and are constant with time ($Q_I \ll Q_U$ such that $Q = Q_U = Q_D$) (d) the upgradient concentration is $C = 0$, and (e) the infiltrating groundwater and groundwater flowing through the aquifer mix completely. Using these constraints and a simple mass balance approach, we can estimate the concentration downgradient of the NAPL contamination, C_D, as $C_D = C_I (Q_I/Q_D)$.

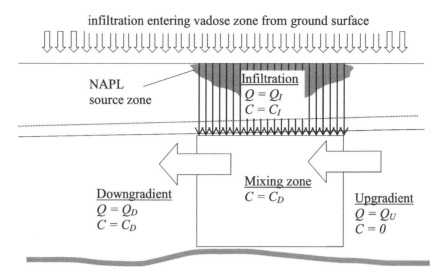

Figure 4.26. Schematic illustration of a simple model for estimating groundwater concentration downgradient of water infiltrating through a NAPL source in the vadose zone.

Accordingly, the concentration in the groundwater is directly related to the infiltration rate, Q_I, implying that temporal variations in recharge will have a significant impact on groundwater concentrations. Of course, temporal variations in Q_U also could be responsible for variations in C_D could be due to variation in Q_U.

This relationship between groundwater concentration and recharge is suggested in data collected from a site contaminated with coal tar in Michigan, USA. Coal tar is a multi-component NAPL originating as a residue from coal gasification operations, and is commonplace throughout the U.S. and Europe. The most hazardous organic compounds associated with coal tar are polynuclear aromatic hydrocarbons (PAHs). The site layout is shown in Figure 4.27, where the plan view shows the configuration of the coal tar source and the PAH plume emanating from the source. The cross-sectional view in Figure 4.27 shows that the coal tar is held above the water table, in the vadose zone.

Figure 4.28 shows concentrations of PAH compounds in groundwater samples taken quarterly from three monitoring wells at a site. The monitoring wells are located within 5 meters down-gradient of areas known to contain coal tar as NAPL trapped in a 2.2-meter thick vadose zone overlying the shallow (4.1 meter thick) aquifer. The sampling results show that there is a pronounced, yearly peak in dissolved PAH concentrations in the late spring, coinciding with the regional snowmelt season, when recharge to the groundwater is the greatest.

The sampling results also show that the degree of the peak concentrations varies from year to year. Snowfall results indicate that 1997–1998 and 1998–1999 total

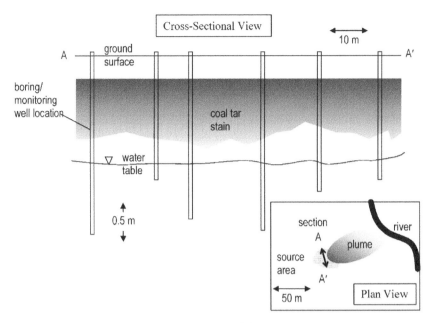

Figure 4.27. Description of coal tar-contaminated site.

snowfalls were almost twice as much as in 1999–2000. Changes in yearly recharge rates related to snowfall variations may explain the year to year variations. Temporal fluctuations in dissolved concentrations also are linked with the sequential raising and lowering of the water table. For example, increases in the water table height may lead to submergence of NAPL previously trapped in the vadose zone. The increased contact between groundwater and NAPL in the saturated zone could lead to increases in dissolution rates and higher concentrations in the surrounding groundwater.

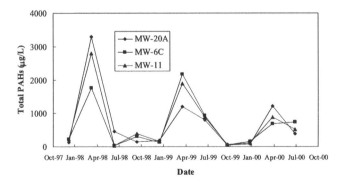

Figure 4.28. Results of periodic sampling of groundwater at a coal tar site.

4.2.2   Upgradient Occurrence of Dissolved NAPL Components via Gas Phase Transport

Once the location of the NAPL source zone is known, we do not expect to find dissolved NAPL in the groundwater upstream of the source zone. However, in many cases, significant concentrations of dissolved NAPL components are found upgradient of the NAPL source zone. Figure 4.29 depicts benzene concentrations and the location of the NAPL source at a gasoline-contaminated site in northern Wisconsin, USA. The benzene plume is not only elongated in the in the downgradient direction from the gasoline source, but also upgradient. No additional NAPL sources have been identified upgradient of the site.

These observations can be explained by diffusive transport of volatilized NAPL components in the gas phase, followed by dissolution of volatilized components into the groundwater. Figure 4.30 shows a schematic illustration of a LNAPL release and subsequent occupation of LNAPL in the vadose zone. A dissolved NAPL source is created from NAPL dissolving in the vadose zone, followed by vertical migration of dissolved NAPL to the water table and by NAPL dissolving at the NAPL-groundwater interface. The dissolved NAPL is transported downgradient by groundwater flow in the saturated zone. In addition, the volatile components of the LNAPL will diffuse from the LNAPL trapped in the vadose zone, creating a vapor

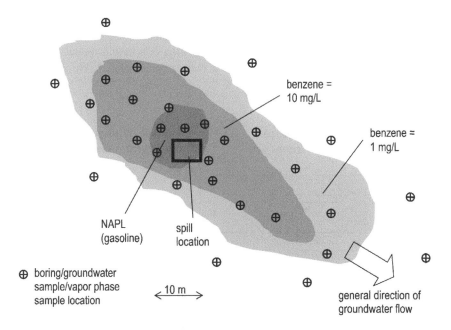

Figure 4.29. Description of gasoline spill site in Northern Wisconsin where significant groundwater concentrations are found upgradient of NAPL source.

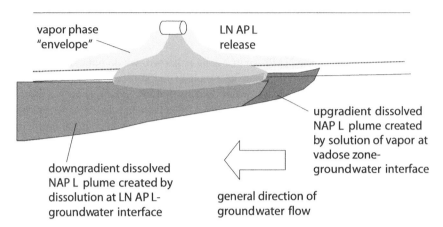

Figure 4.30. Illustration explaining presence of dissolved NAPL components upgradient of NAPL source.

phase "envelope." Diffusion in the gas phase will occur in all directions, including upgradient of the NAPL, as described in Section 2.3.5. As shown in Figure 4.30, the diffused vapor phase NAPL components can partition, or dissolve into the groundwater at the vadose zone-groundwater interface, including at locations upgradient of the NAPL source. Infiltrating water may dissolve some of the vapor phase and carry it down to the water table, upgradient of the source.

Figure 4.31 schematically describes a simplified conceptual model of these processes, from which we can develop a mathematical model. We apply the

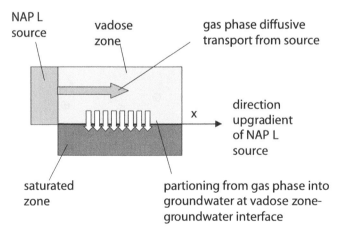

Figure 4.31. Conceptual model of diffusive transport and partitioning processes.

one-dimensional form of Fick's law to describe the diffusive transport from the NAPL source into the vadose zone, as follows:

$$\frac{\partial C_a^i}{\partial t} = D_a^{i\,eff} \frac{\partial^2 C_a^i}{\partial x^2} \qquad (4.37)$$

where x is the horizontal distance, t is time, and $D_a^{i\,eff}$ is the effective diffusivity of chemical component i in the gas phase. Given (a) an initial condition of $C_a^i = 0$ for all x, (b) an infinite boundary in the positive x-direction and (c) a boundary condition of $C_a^i = (C_a^i)_0$ at $x = 0$, a we can solve Equation (4.37) to obtain:

$$\frac{C_a^i}{(C_a^i)_0} = erfc\left(\frac{x}{2\sqrt{D_a^{i\,eff}\,t}}\right) \qquad (4.38)$$

where erfc is the complementary error function and $(C_a^i)_0$ represents the concentration of a LNAPL component in the gas phase at the NAPL source boundary. We can also write this equation in terms of partial pressures

$$\frac{p_a^i}{(p_a^i)_0} = erfc\left(\frac{x}{2\sqrt{D_a^{i\,eff}\,t}}\right) \qquad (4.39)$$

where $(p_a^i)_0$ is determined by $(p_a^i)_0 = (\chi_o^i)_0\, p_{vap}^i$

We then apply Henry's law at the vadose zone-groundwater interface to determine the concentration in the groundwater at the interface, C_w^i:

$$C_w^i = \frac{p_a^i}{H^i} M_{wt}^i \qquad (4.40)$$

where M_{wt}^i is the molecular weight of component i. The conceptual model shown in Figure 4.31 and Equations (4.39) and (4.40) are applied in the following example.

Example: Consider the transport of benzene from multicomponent NAPL, where the benzene is present at a mole fraction of $\chi_o^i = 0.01$. We determine the effective diffusivity as

$$D_a^{i\,eff} = \tau_a D_a^i = \frac{(\phi S_a^i)^{10/3}}{\phi^2} D_a^i \qquad (4.41)$$

where D_a^i is the chemical diffusion coefficient in free gas, S_a^i is air saturation, and τ_a is the tortuosity factor in the gas phase [Millington and Quirk, 1961].

Given (a) the diffusivity of benzene in gas phase as $D_a^i = 8.8 \times 10^{-2}$ cm²/s, (b) the Henry's constant as $H^i = 5.56 \times 10^{-3}$ atm-m³/mol @ 25°C, (c) the vapor pressure of benzene as $p_{vap}^i = 0.1252$ atm @ 25°C, (d) an air saturation of 0.33, and (e) a porosity of 0.3, we can determine the gas phase partial pressures in the vadose zone as a function of distance and time using Equation (4.39) and the corresponding groundwater concentrations using Equation (4.40).

Figure 4.32(a) shows the gas phase concentrations as a function of distance from the NAPL source at various times since the NAPL source was released into the vadose zone. In this figure, we see that significant gas phase transport away from the source within one to 10 years. The effect of the vapor phase transport on underlying groundwater is shown in Figure 4.32(b), where benzene groundwater concentrations exceed 1 mg/L at a distance of 20 to 50 meters upgradient from the source after about ten years. Of course these results are specific to the chosen parameters (mole fraction in the NAPL, air saturation, etc.), but in general we recognize that the traveled distance of a diffusive front varies with time as $x = \sqrt{D_a^{i^{eff}}\, t}$.

A spreadsheet file (gasphasediffusion.xls) that can be used to make calculations of benzene transport from a NAPL source in the gas phase and partitioning into the aqueous phase is found on the accompanying CD.

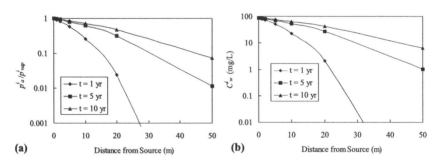

Figure 4.32. (a) Benzene partial pressures normalized by vapor pressure at various times since the NAPL was released into the subsurface and (b) groundwater concentrations at the vadose zone-groundwater interface.

Retardation can occur in the vadose zone (see Section 2.3.4), which will have the effect of slowing the diffusive transport in the gas phase and decreasing the corresponding upgradient concentrations in the groundwater. Using Equation (2.46), we can calculate a retardation factor, R_a^i. The diffusive transport is now described as

$$\frac{p_a^i}{(p_a^i)_0} = erfc\left(\frac{x}{2\sqrt{D_a^{i^{eff}} t / R_a^i}}\right) \tag{4.42}$$

and the traveled distance is described as $x = \sqrt{D_a^{i^{eff}} t / R_a^i}$. The impact of retardation is illustrated in the following example.

Example: Consider the transport of benzene from multicomponent NAPL as described in Example, but allow for the effects of retardation. Using a bulk density of $\rho_b = 1.65$ gm/cm^3, a solid-water partition constant of $K_d = 0.001$ cm^3/gm, and the parameter values we have listed previously, we obtain $R_a^i = 8.09$. In Figure 4.33, we have plotted the groundwater concentrations, taking into account the retardation factor. We see that the concentrations have been reduced considerably compared to Figure 4.32, but still reach 1 mg/L after 20 years.

A spreadsheet file (gasphasediffusion_retard.xls) that can be used to calculate transport of benzene from NAPL source in gas phase and its partitioning into aqueous phase is found on the accompanying CD.

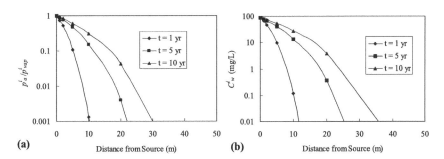

(a)

(b)

Figure 4.33. Effects of retardation: (a) benzene partial pressures normalized by vapor pressure at various times since the NAPL was released into the subsurface and (b) groundwater concentrations at the vadose zone-groundwater interface.

5

Remediation

5.1 REMEDIATION METHODOLOGIES

Over the last few decades, dozens of technologies for remediating NAPL-contaminated sites have been developed and applied in the field, yet remediation of NAPL-contaminated sites is still problematic. For most sites, the selection of the appropriate remediation technology is not straightforward, and design guidelines for remediation technologies are not well established. Although several innovative technologies have shown promise, only a few standard technologies are widely applied in the field. In the following sections, we concentrate on observations and phenomena associated with the application of the following, most frequently applied technologies.

- Hydraulic removal of LNAPL: extraction of LNAPLs via wells or trenches, based on the manipulation of LNAPL and/or water gradients to drive mobile NAPL towards extraction points (Section 5.2).
- Pump-and-treat: Dissolution of NAPL components and extraction of ground-water containing dissolved NAPL components via extraction wells (Section 5.3).
- Soil vapor extraction: Volatilization of NAPL components and extraction of vapor phase containing volatilized NAPL components via vapor-phase extraction wells (Section 5.4)

First, we briefly describe a number of in-situ NAPL remediation technologies. These technologies rely on containment of the pollution (capping, hydraulic isolation, soil vitrification, and solidification), removal of pollutants (excavation, pump and treat, soil vapor extraction, hydraulic removal), inducing phase transfer (air sparging), intercepting pollutants (reactive permeable barrier, funnel and gate), changing the physico-chemical properties of the NAPL (chemical flushing), inducing phase changes (thermal enhancement), chemical degradation of NAPL (oxidation), or chemical or biochemical degradation of NAPL components dissolved in groundwater (oxidation, biodegradation). A general description of these technologies is given in Table 5.1, along with references where more details can be found. In addition to these references, Table 5.2 gives a list of web sites where information on applications of these and other remediation technologies can be found.

Soil and Groundwater Contamination: Nonaqueous Phase Liquids
Water Resources Monograph 17
Copyright 2005 by the American Geophysical Union
10.1029/17WM05

TABLE 5.1.   List of selected remediation technologies.

Technology	Basic principles	References
Containment • *Capping* • *Hydrodynamic isolation* • *Solidification* • *Vitrification*	In containment methods, the hydraulic contact between the polluted zone and the rest of the soil and groundwater system is inhibited. The spread of NAPL and dissolution of its components are prevented or significantly slowed down. *Capping* involves placing a cover over contaminated material to prevent escape of harmful vapors and stop rainwater from seeping through the polluted zone. *Hydrodynamic isolation* of the polluted zone is achieved by modifying the local flow regime through a strategic placement of pumping and injection wells. *Solidification* involves mixing polluted soil with materials such as cement to harden. *Vitrification* is a process that turns the soil into a solid block of glasslike material. Electricity is used to create the heat needed to melt the soil.	EPA [1991] EPA [1999a] Rumer and Ryan [1995] NRC [1997] Freeman and Harris [1995] Bogacki and Daniels [1989]
Removal • *Excavation* • *Pump and treat*	In removal methods, soil and/or ground-water is removed and treated on site. *Excavation* is simply the digging up of polluted soil. In *pump and treat,* polluted groundwater is brought up to the surface through pumping. It relies on advection and dissolution processes.	EPA [1991] EPA [1997] EPA [1999b] EPA [1999c] Wilson [1995]
Intercepting Pollutants • *Reactive permeable barrier* • *Funnel and Gate*	In these methods, polluted groundwater is intercepted downstream of the pollution site where, in a controlled zone, the pollutants are removed by means of chemical and biological methods. A *permeable reactive barrier (PRB)* is created by building a long narrow trench in the path of the polluted	Naftz et al. [2002] Wickramanayake et al. [2000] Gavaskar et al. [1998] Spooner et al. [1985]

TABLE 5.1. Continued.

Technology	Basic principles	References
	groundwater and filling it with a material that can clean up dissolved NAPL components. This barrier may have strong adsorptive properties, oxidating properties, ion exchange properties and/or contain bacteria that biodegrade NAPL components. A *funnel-and-gate* system is basically similar to PRB, except that by guiding walls (funnel) along the groundwater flow direction, the groundwater is forced to pass through the permeable reactive zone (gate).	
Inducing Phase Transfer • *Soil vapor extraction* • *Air sparging*	*Soil vapor extraction* (SVE) removes volatile NAPL components from the unsaturated zone through extraction wells. By creating vacuum in the wells, polluted air is withdrawn from unsaturated zone and volatilization is promoted. If biodegradation is also promoted actively, it is called bioventing. *Air sparging* involves injecting air below the groundwater table. The air strips soil and groundwater volatile NAPL components from and carries them to the unsaturated zone or the surface where they are collected by a collection system. The air also replenishes groundwater oxygen and promotes bio- degradation of dissolved NAPL components. Air sparging is often used together with soil vapor extraction.	Wilson [1995] Hinchee et al. [1995] API [1996]
Chemical Flushing • *Surfactant-enhanced aquifer remediation*	Mobilization: Injection of chemical agent dissolved in water reduces interfacial tension between NAPL and water. Reduced interfacial correspondingly reduces capillary pressure, such that imposition of a hydraulic gradient on	Falta et al. [1999] Jackson [2003] Londergan et al. [2001] Rao et al. [1997]

TABLE 5.1.   Continued.

Technology	Basic principles	References
• *Solvent* *flushing*	NAPL can overcome resisting capillary pressure. Solubilization: Injection of chemical agent dissolved in water increases solubility of NAPL components dissolved in groundwater. Increased solubility correspondingly increases the efficiency of extraction of ground-water containing dissolved NAPL components.	Sabatini et al. [1996] Simpkin et al. [1999]

Chemical Oxidation
• *Potassium*
   permanganate
• *Hydrogen*
   peroxide +
   Fenton's
   reagent
• *Ozone*
	A chemical agent is introduced that degrades NAPL components dissolved in groundwater. Degradation of dissolved NAPL components near NAPL-water interface results in enhanced rates of mass transfer, increasing the efficiency of extraction of groundwater containing dissolved NAPL components. In some cases, the chemical oxidant will degrade the NAPL directly, in addition to reacting with NAPL dissolved in groundwater.	Gates and Siegrist [1995]
		Nelson et al. [2001]
		Seol et al. [2001]
		Wickramanayake et al. [2000]

Thermally-Enhanced Extraction
• *Steam injection*
• *Electrical*
   resistance
   heating
	Steam injection at the periphery of the NAPL-contaminated area heats the subsurface and drives NAPL towards liquid extraction wells. Steam injection also vaporizes volatile compounds in NAPL and drives vaporized NAPL components towards vapor extraction wells. Electrical heating of subsurface vaporizes volatile NAPL components; vaporized components are collected via vapor extraction wells.	Wickramanayake and Hinchee [1998]

Chemical Oxidation/ Thermal Enhancement
• *Hydrous*
   pyrolysis
   oxidation
	Steam and air are injected into the NAPL contaminated area, creating a heated, oxygenated zone. The injection is stopped and the steam condenses and groundwater containing dissolved NAPL components returns to the heated zone. The groundwater mixes	Knauss et al. [1997]
		http://apps.em.doe.
		gov/ost/pubs/itsrs/
		itsr1519.pdf

TABLE 5.1.   Continued.

Technology	Basic principles	References
	with the condensed steam and oxygen, which degrades the NAPL components dissolved in water. The heat enhances the rate of oxidation and the solubility of NAPL components.	
Bioremediation • *Engineered Bioreme- diation* • *Natural Attenuation*	Biochemical reactions degrade NAPL components dissolved in groundwater. Degradation of dissolved NAPL components near NAPL-water interface results in enhanced rates of mass transfer, increasing the effective rate of dissolution and the efficiency of extraction of groundwater containing dissolved NAPL components.	Seagren et al. [2002]

TABLE 5.2.   Limited list of web sites with information on NAPL remediation technologies.

Web Site	Web Address
Environmental Security Technology Certification Program	http://www.estcp.org/
Federal Remediation Technologies Roundtable	http://www.frtr.gov/
Groundwater Remediation Technologies Analysis Center	http://www.gwrtac.org/
Interstate Technology and Regulatory Cooperation	http://www.itrcweb.org/ common/default.asp
U.S. Department of Defense Strategic Environmental Research and Development Program	http://www.serdp.org/
U.S. EPA Remediation and Characterization Innovative Technologies (REACH-IT)	http://www.epareachit.org/
U.S. EPA Remediation Technologies Development Forum	http://www.rtdf.org/
U.S. EPA Superfund Innovative Technology Evaluation Program	http://www.epa.gov/ORD/SITE/
U.S. EPA Technology Innovation Program Hazardous Waste Cleanup Information (CLU-IN)	http://www.clu-in.org/
U.S. National Institute of Environmental Health Sciences Superfund Basic Research Program	http://www-apps.niehs.nih.gov/ sbrp/index.cfm

5.2   HYDRAULIC REMOVAL OF LNAPL

5.2.1   LNAPL Trapping as Free Product Migrates Towards Extraction Well

Summary: Two processes can contribute to the eventual trapping of LNAPL during LNAPL recovery efforts: (a) reduction of LNAPL saturations to a level where the LNAPL exists in a residual, disconnected state and (b) reduction of LNAPL saturations, and corresponding LNAPL hydraulic conductivities, to the point where LNAPL flow rates decrease to a negligible level.

Hydraulic removal of LNAPL relies on the extraction of LNAPLs via wells or trenches, based on the manipulation of LNAPL and/or water gradients to drive free LNAPL towards extraction points. Extraction wells or trenches are installed and operated to remove lighter-than-water NAPL (LNAPL) from the top of the water table. However, extraction efforts are not capable of removing all of the LNAPL from the aquifer. Table 5.3 gives results from several case studies, where the amount of LNAPL remaining after extraction is indicated. Table 5.3 shows that 64-83% of the LNAPL is not recovered, which is typical for LNAPL extraction operations. In general, the primary benefit that can be achieved with this technology is limiting future migration of LNAPL, due to reduced LNAPL mobility or transmissivity.

TABLE 5.3.   Results from selected LNAPL recovery efforts.

Location	Volume of Spill (m³)	LNAPL	Extraction System	Volume Recovered (m³)	Percent Remaining
Port Huenemene, California	200	jet fuel	conventional wells	44	78
Mahwah, New Jersey	500	fuel oil	dual-phase wells	170	66
Phoenix, Arizona	160	gasoline	trenches and conventional wells	46	71
Tacoma, Washington	680	gasoline	conventional wells	245	64
Jacksonville, North Carolina	850	jet fuel	conventional and dual phase wells	145	83
Port Huron, Michigan	300	gasoline	conventional wells	75	75
New Haven, Connecticut	130	fuel oil	trenches	26	80
East Saint Louis, Illinois	300	gasoline	conventional wells	84	72

Free LNAPL is defined here as LNAPL that exists at a positive pressure (a pressure greater than atmospheric pressure). The positive pressure condition implies that the LNAPL is present at saturations higher than residual, that is, the LNAPL is found as an interconnected liquid within the pore space, along with varying amounts of water and air. When a hydraulic gradient is applied to the free LNAPL, the gradient is "felt" by the entire volume of LNAPL, as long as the LNAPL sat-uration is above the residual level.

In order to understand the behavior of free LNAPL that is flowing to an extraction well or trench, it is useful to examine the process of LNAPL emplacement in the vadose zone. Once an LNAPL is released into the vadose zone, as a result of a surface spill, underground tank leak, or other means, it will migrate vertically through the vadose zone. If a sufficient release volume is present, the LNAPL will reach the top of the capillary fringe. Figure 5.1 illustrates the LNAPL plume configuration as the LNAPL migrates downward from its point of release to the capillary fringe. In this scenario, the LNAPL is released into the subsurface from an underground storage tank and is migrating vertically due to gravity and capillary forces and horizontally due to capillary forces. The top set of figures illustrates the developing LNAPL plume. The middle set of figures shows the saturation of

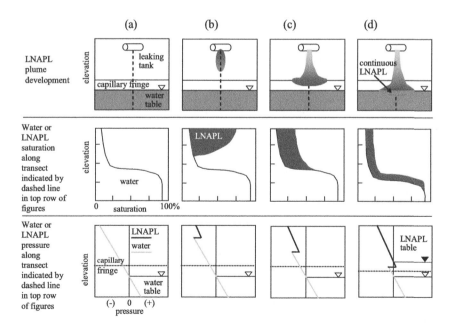

Figure 5.1. Schematic illustration of LNAPL plume configuration, saturation distribution and pressure distribution, (a) before the NAPL release, (b) during the migration of the LNAPL plume, (c) during development of an LNAPL capillary fringe, and (d) after development of an LNAPL table.

each fluid phase with depth, along the vertical transect designated with a dashed line in the upper figures. The bottom set of figures indicates the pressure distribution with depth along the same vertical transect.

Figure 5.1(a) shows the configuration of the capillary fringe before the spill is released, when only water and air are present. As the LNAPL plume is released and moves downward, it will displace some of the capillary water held in the vadose zone (Figure 5.1(b)). During the initial stages of accumulation, the LNAPL will be under negative pressure (less than atmospheric pressure). Eventually, the LNAPL reaches the capillary fringe and the water table, and begins to accumulate on top of the water table, as shown in Figure 5.1(c). When a sufficient volume of LNAPL has accumulated on top of the water table, an "LNAPL table" will form. The LNAPL at the bottom of the LNAPL table will now have a positive pressure, as shown in Figure 5.1(d). In Figure 5.1, the slopes of the LNAPL and water elevation vs. pressure lines may appear to be the equal; the LNAPL elevation vs. pressure line is actually steeper than that for the water, since the LNAPL is lighter than water.

Although, by definition, the LNAPL is lighter than water, the LNAPL will accumulate below the level of the surrounding water table, as indicated in Figure 5.1(d). The apparent "sinking" of the LNAPL below the water table is explained by Archimedes' principle: a floating object will displace a volume of water that has a weight equal to the weight of the object that is floating. Figure 5.1(d) also labels the LNAPL accumulating on top of and below the water table as being "continuous." The continuity of the LNAPL is significant because the LNAPL body can be displaced by a hydraulic gradient as long as the fluid is continuously linked throughout the network of proe spaces.

The migration process described in Figure 5.1 could take a matter of several hours to months, depending on the volume of the LNAPL release, the position of the point of release relative to the top of the capillary fringe, properties of the porous media (hydraulic conductivity and capillary properties), and the LNAPL fluid properties (density and viscosity).

In Figure 5.2, a sequence of events that might occur during LNAPL recovery is illustrated. Again we point out that, in Figure 5.2, the LNAPL elevation vs. pressure line is actually steeper than that for the water, since the LNAPL is lighter than water. In Figure 5.2, an extraction well has been installed in the center of the LNAPL plume. The extraction well is screened through the LNAPL plume. Since the LNAPL pressure head is positive in the continuous LNAPL, the LNAPL will flow into the extraction well, even without pumping. Pumping in the extraction well will lower the fluid level in the well, producing an LNAPL hydraulic gradient and LNAPL migration towards the well. The continuous LNAPL is gradually extracted by the well, as shown in Figure 5.2(b) and Figure 5.2(c).

Figure 5.2(d) indicates that, as the LNAPL saturation in the zone above the continuous LNAPL zone is reduced, the LNAPL will be replaced in the pore space by water. The LNAPL above the continuous LNAPL zone also can be removed by the extraction well, as long as the following conditions hold: (1) there

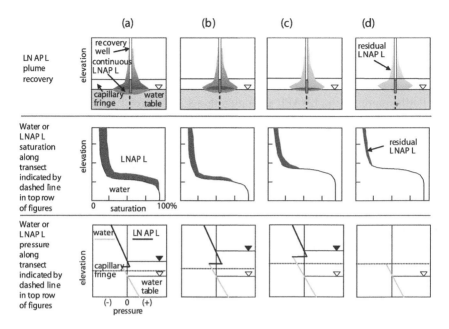

Figure 5.2. Schematic illustration of NAPL plume configuration, saturation distribution, and pressure distribution, (a) before LNAPL extraction, (b)-(c) during LNAPL extraction, (d) after LNAPL extraction has ceased.

is a hydraulic connection between the LNAPL in the well and the LNAPL in the porous medium and (2) a positive hydraulic gradient is maintained, that is, if the head in the aquifer is greater than the head in the well. The equivalent water head is defined as

$$h_o = z_{ow} + D_o \frac{\rho_o}{\rho_w} \qquad (5.1)$$

where z_{ow} is the elevation of the LNAPL-water interface and D_o is the thickness of the LNAPL layer in the porous medium, defined in Section 4.1.1.

Practically speaking, however, flow to the well will diminish to almost negligible levels as the LNAPL saturation drops, say below 20%. Darcy's law for an LNAPL flowing in the horizontal direction is given as:

$$q_o = -\frac{\rho_w}{\rho_o} K_o \frac{dh_o}{dl} \qquad (5.2)$$

where q_o is the LNAPL specific discharge, K_o is the LNAPL hydraulic conductivity, and dh_o/dl is the LNAPL gradient. The ratio of densities is needed here because h_o is the water-equivalent head. As shown in Figure 2.12, when the

LNAPL saturation is low, the LNAPL relative permeability, and correspondingly, the LNAPL hydraulic conductivity, decreases sharply. A sharply decreased LNAPL hydraulic conductivity will produce a diminishing LNAPL specific discharge, resulting in an LNAPL flow rate that approaches zero towards the extraction well.

Furthermore, as the LNAPL saturation drops in the free LNAPL zone, water will replace the LNAPL in the pore spaces. Above capillary fringe, air may also enter the pores. Accompanying the replacement of LNAPL by water and air into the pores, LNAPL ganglia will be cut off by the migrating water, resulting in residual, trapped LNAPL. In Figure 5.2(b), (c), and (d), a zone of residual LNAPL is shown to remain after the mobile LNAPL is removed. This process is further illustrated in Figure 5.3, where the LNAPL residing on the capillary fringe is idealized as a layer of uniform thickness. As the LNAPL in the layer migrates towards the extraction well, the saturation in the layer is reduced, eventually leaving behind residual LNAPL. Figure 5.3 also shows the path traced along the capillary pressure-saturation curve at a fixed location within the LNAPL layer. The capillary pressure is approaches zero as the LNAPL saturation decreases and the water saturation increases. The water saturation never reaches 100% due to

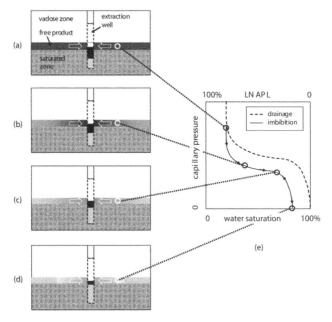

Figure 5.3. (a)-(d) Changes in saturation in a layer of free LNAPL during LNAPL extraction from a well. Shading qualitatively represents ranges of LNAPL saturation, with the darkest shade indicating high saturations and the lightest shade representing residual saturation. (e) Corresponding capillary pressure-saturation curve. The position on the capillary pressure-saturation drainage curve is given for the corresponding location in (a)-(d).

the presence of the residual LNAPL. The amount of residual LNAPL remaining depends on the pore size distribution and the LNAPL fluid properties.

5.2.2   LNAPL Trapping in the Cone of Depression

Summary: Production of fluids (water and LNAPL) might result in lowering of the water and LNAPL tables and the formation of a smearing zone. LNAPL entrapment is likely to occur during single- or dual-pump well operations. Smearing zone formation may be minimized by careful selection of LNAPL and/or water extraction rates. Vacuum-enhanced recovery reduces smearing and subsequent entrapment. The relation between free-LNAPL recovery techniques and the potential of LNAPL entrapment in smearing zones is discussed.

Extraction of water and LNAPL might result in lowering of the water and LNAPL tables. As a result of the drawdown, a smearing zone develops. If only LNAPL is produced during a recovery operation with a single- or dual-pump well system, the interface between the LNAPL and water forms a cone of depression. Gradually, increasing the water production will increase the LNAPL production rate and eventually cause a drawdown of the water-LNAPL interface and the creation of a smearing zone. The maximum water recovery rate associated with a LNAPL recovery rate resulting in no drawdown of the water-LNAPL interface is called the *maximum zero-smearing* rate. When the water rate exceeds this value, smearing occurs.

Smearing is the process when LNAPL is drawn into previously uncontaminated parts of the aquifer (Figure 5.4). The LNAPL that is drawn down displaces water and is not entrapped during pumping. However, some of the smeared LNAPL will become entrapped and discontinuous below the water table after extraction ceases and water displaces the LNAPL. When assessing the potential for the development of smearing zones during remediation, it is important to recognize the fluid withdrawal features of the various free-LNAPL recovery technologies. The main technologies are trench, skimmer well, single- or dual-pump, and vacuum-enhanced recovery (also known as "bioslurping") systems.

Interceptor trenches, generally placed down gradient from a LNAPL release and perpendicular to the direction of groundwater flow, are primarily used to control the migration of free LNAPL. The rate of flow to the trench is largely determined by the natural hydraulic gradient. Trenches are usually installed up to depths of about 10 m. Advantages of this type of recovery are the continuous lines of interception, suitability for large releases, and, in case of shallow trenches, a straightforward construction. Disadvantages are the limited applicability in low permeability soils, problematic placement near buildings and utilities, and cost. The American Petroleum Institute [1996] recommends using trenches for areas with a shallow water table and where the saturated thickness of the aquifer is small, resulting in ineffective wells.

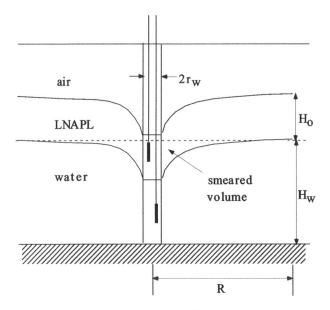

Figure 5.4. Smeared zone formation as a result of drawdown.

Skimmer well free-LNAPL removal systems produce only LNAPL without water and are often used as an initial response action. They are best suited for sites where the LNAPL aerial extent and thickness are relatively small. Skimmer wells are also sometimes used in trenches. Advantages of this method are the limited production of water and low-cost operation and installation. The major limitations are the small area of influence and the lack of imposed hydraulic control to limit migration of the NAPL and/or the dissolved plume. These kinds of wells are used at sites where water table depression does not lead to enhanced LNAPL recovery.

Single- and dual-pump recovery systems use water table manipulations to expand the area of influence. The created cone of depression draws LNAPL to the well and these systems usually produce LNAPL at much larger rates than skimmer wells. Single-pump recovery wells use a single pump intake and typically both water and LNAPL are produced. In dual-pump wells, one pump is used to create a cone of depression and the other to withdraw the LNAPL that is moving into the well. For both systems, the LNAPL production usually increases when the drawdown increases. However, this leads to LNAPL smearing during the pumping procedure and entrapped LNAPL after the production has stopped. Advantages of this method are versatility, relatively low cost, possibility of use at congested sites, and ability to offer hydraulic control. Limitations are maintenance problems (including biofouling), required movement of well inlets when fluid elevations are changing, use of a separator (for single-pump wells), treatment of produced water, and the formation of a smearing zone. Dual-pump wells are preferred over single-pump wells when water table depression will considerably increase recovery and/or hydraulic control of the LNAPL body, and when separation of large volumes of LNAPL/water mixtures is impractical.

Vacuum-enhanced recovery methods, such as bioslurping, are used in combination with, and in order to increase the performance of, single- and dual-pump systems. The applied vacuum increases the driving force and subsequent fluid movement towards a well. These systems force air in the unsaturated zone to move towards the well. The moving air may volatilize residual organics, strip contaminants from the aqueous phase, and transport volatile organics to the well. The main advantages of vacuum-enhanced systems are:

(a) smearing can be avoided,
(b) contaminants can be removed from the vadose zone through air movement,
(c) they are effective in low-permeability zones,
(d) application of a vacuum to a well increases the hydraulic gradient and, as a result, the capture, and
(e) they remain effective when LNAPL layer thickness is small.

Drawbacks of this method are the potential high concentration of dissolved components in wastewater and maintenance costs. Based on the characteristics of the free-LNAPL recovery methods, single- or dual-pump wells without vacuum enhancement typically create smearing zones.

The physical principles of vacuum-enhanced extraction are explained in Figure 5.5, following Nyer et al. [1996]. The basic idea behind vacuum-enhanced recovery is to increase the hydraulic gradient and the capture zone beyond that can be achieved by pumping alone. Vacuum-enhanced recovery combines vapor extraction (Figure 5.5(a)) with groundwater pumping (Figure 5.5(b)) into an integrated system (Figure 5.5(c)). In Figure 5.5a, an applied vacuum to a well yields a cone of impression as a result of the negative pressure. In this example, the water table rise in the well is 1 m. The water levels from pumping alone are shown in Figure 5.5(b). For a certain pumping rate, the drawdown at the well is 2 m. Combining the effects of pumping and the effects of vacuum extraction (Figure 5.5(c)), yields a net drawdown of only 1 m. This is the value that one would measure in the well. However, the effective drawdown head, which is a combination of the pressure and liquid gradient, is still 2 m. In addition, the capture zones for the situation depicted in Figure 5.5(b) and Figure 5.5(c) are the same. The principles behind Figure 5.5 are applied in the field when attempting keep the smeared zone to a minimum by balancing the upconing through vacuum pumping and drawdown due to water pumping.

A special case of vacuum-enhanced recoveryis bioslurping, where airflow in the vadose zone not only causes the removal of contaminants through volatilization, but it also promotes biodegradation. If applied properly, bioslurping techniques, as well as conventional skimmer, create a very small smearing, whereas dual-pump wells can create a substantial smearing zone.

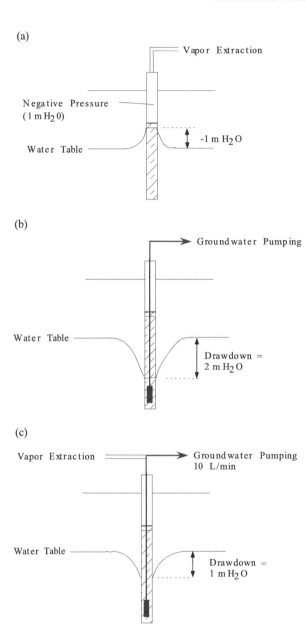

Figure 5.5. Effect of (a) vapor extraction, (b) groundwater pumping, and (c) combination of the two on drawdown (after Nyer et al. [1996]).

A simple formula to estimate the volume of a smearing zone, V_{sm}, can be derived for well operations where the drawdown is relatively small compared to the saturated thickness of the aquifer. In that case, the well-known Dupuit assumptions [Bear, 1972] apply; i.e., one may assume that the streamlines near a free surface boundary are horizontal and the slopes of the line of seepage and the hydraulic gradients are equal. For unconfined aquifers, the Dupuit assumptions lead to the following relation between water heads (h) and distances to a pumping well (r) for steady-state conditions at two arbitrary locations a and b:

$$h_a^2 - h_b^2 = \frac{Q_w}{\pi K_w} \ln\left(\frac{r_a}{r_b}\right) \tag{5.3}$$

A common rule-of-thumb is that when the drawdown in the well does not exceed one-half of the original aquifer thickness, the use of Equation (5.3) leads to reasonable results, even when h_b is measured at the well periphery where $r_b = r_w$ or when h_a is equal to the water head at the well capture zone boundary, h_c. Denoting the capture zone radius as r_c, Equation (5.3) becomes:

$$h_c^2 - h_b^2 = \frac{Q_w}{\pi K_w} \ln\left(\frac{r_c}{r_b}\right) \tag{5.4}$$

Solving Equation (5.4) for h_b yields:

$$h_b(r_b) = \left\{ h_c^2 - \left[\frac{Q_w}{\pi K_w} \ln\left(\frac{r_c}{r_b}\right) \right] \right\}^{0.5} \quad \text{for} \quad r_w \leq r_b \leq r_c \tag{5.5}$$

When the upper and lower boundaries for a smeared zone are given by h_c and h_w, respectively, the smeared volume can be written as:

$$V_{sm} = 2\pi \int_{r_w}^{r_c} \left(h - \left\{ h^2 - \left[\frac{Q_w}{\pi K_w} \ln\left(\frac{r_c}{r}\right) \right] \right\}^{0.5} \right) r\, dr \tag{5.6}$$

Example plots showing V_{sm}, computed with Equation (5.6), as a function of r_c for two values of K_w are presented in Figure 5.6. In this hypothetical example, $r_w = 0.05$ m, and $Q_w = 20$ m^3/day. The figure shows that for a lower-permeability aquifer, the drawdown and the smeared volumes are larger.

A set of analytical expressions developed by Johns et al. [2003] provides tools to estimate LNAPL and water extraction rates considering smearing behavior in single- and dual-pump recovery well systems. The assumptions behind the equations include:

(a) fluids are incompressible and have constant viscosities,
(b) wells produce LNAPL and water at constant rates,

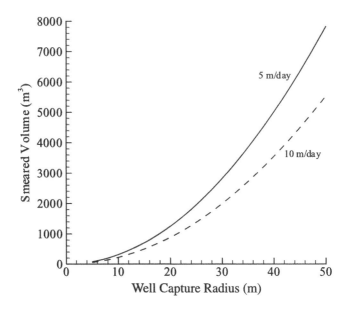

Figure 5.6. Smeared volume (in m^3) as a function of well capture radius for $K_w = 5$ m/day and $K_w = 10$ m/day.

(c) flow is radial toward the well,
(d) fluids are in vertical equilibrium,
(e) the aquifer is homogeneous and has a constant thickness,
(f) capillary pressure between fluids is negligible, and
(g) the aquifer material has a relatively large hydraulic conductivity.

The equations derived by Johns et al. [2003] allow the user to develop operational windows for single-dual phase pumping based on fluid and porous medium properties and certain design criteria. The operational windows plot water versus LNAPL extraction rates and include zones where no smearing occurs. Based on their analytical analysis for several cases, Johns et al. [2003] make the following conclusions. (a) A skimmer well avoids smearing of the LNAPL below the water table, but the recovery rate is relatively small. (b) Pumping aqueous phase with a single- or dual pump well may increase the recovery by a factor of two or three over the skimmer rate while avoiding smearing. Increases beyond a factor of two or three may be achieved by increasing the aqueous phase pumping rate, but this will result in the formation of a smearing zone. However, creating a smearing zone might be acceptable if a smearing zone is already in place due to fluctuating water table behavior.

5.2.3   Decrease in Free LNAPL Recovery Rates as a Function of Time

Summary: During field operations of a LNAPL recovery unit, LNAPL extrac-tion invariably reduces over time. The main reasons for this decline are decreasing amounts of mobile LNAPL, decreasing LNAPL relative permeabil-ity, and poorly designed and operated wells. Examples of analytical solutions to compute recovery times and LNAPL extraction rates are discussed in the context of the imposed assumptions. The analytical solutions apply to rela-tively simple cases. For more complex scenarios, multi-fluid flow numerical models might be necessary.

Hydraulic recovery is the result of an induced hydraulic gradient forcing LNAPL to flow to wells or trenches. The rate of flow is a strong function of the transmissivity of the LNAPL in the vicinity of the recovery unit. The LNAPL transmissivity is the product of the porous medium permeability and the LNAPL relative permeability, which in turn is a function of saturation, and the thickness of the LNAPL. When LNAPL is removed, the saturation and thickness of the mobile LNAPL fraction decreases, resulting in a reduced recovery rate. If this effect occurs near a well, through overly aggressive pumping, movement of LNAPL from further distances to the well is negatively affected.

Most, if not all, LNAPL field recovery operations have shown an asymptotic LNAPL response. In a case study of diesel fuel removal through recovery wells from a dune sand in the USA, a rapidly diminished efficiency was observed over a four-year period [Huntley and Beckett, 2002a]. The fraction of LNAPL in the recovered fluids decreased from an initial maximum of 0.01 to less than 0.0003, with a projected-averaged fraction of just 0.002. A graph of the LNAPL fraction in total fluid and the cumulative LNAPL recovery as a function of time is shown in Figure 5.7. Core analysis and site excavation showed that large volumes of the diesel oil remained in the soil, even when virtually no LNAPL was recovered anymore. It was estimated that, based on a total solubility of 15 mg/l, the remain-ing LNAPL would reside at the site for at least several thousand years.

Several analytical expressions based on simplified assumptions have been derived to compute LNAPL recovery rates using trenches, skimmer wells, and single- and dual-pump wells with or without vacuum enhancement. Input para-meters for the computations usually include LNAPL fluid properties, porous medium properties, the LNAPL well thickness (H_o), and relations between H_o and the LNAPL specific volume, V_o (see Section 4.1.1). Several of these recovery expressions are listed in Charbeneau et al. [1999, 2000]. For instance, a simplified expression for the LNAPL recovery rate, Q_o, from a single-pump well is given as

$$Q_o = \left(\frac{\rho_{ro}\beta^2 Q_w}{\mu_{ro}\phi^2 H_w} \right) \frac{(H_o - \alpha)^2}{H_o} \tag{5.7}$$

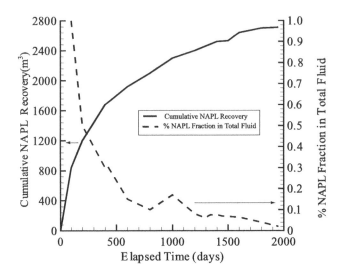

Figure 5.7. Fraction LNAPL in recovered fluids and cumulative LNAPL recovery (m^3) for a dune sand site (after Huntley and Beckett [2002a]). Reprinted with permission from American Petroleum Institute (API).

where Q_w is the water recovery rate, ρ_{ro} is the LNAPL relative density, μ_{ro} is the LNAPL relative viscosity, ϕ is the porosity, H_w is the saturated water thickness at the well radius of influence. Parameters α and β are porous-medium-specific and appear in Equation (4.23), which is an approximate linear relation between well LNAPL well thickness, H_o, and LNAPL specific volume V_o.

Examples of LNAPL recovery rates as a function of well LNAPL thickness (H_o) for single-pump wells (based on Equation (5.7)) are shown in Figure 5.8 for three different water extraction rates, 1, 5, and 10 L/min, in a sandy loam soil. The porosity, LNAPL relative density, LNAPL relative viscosity, and saturated water thickness were assumed to be 0.4, 0.7, 2.0, and 5.0 m, respectively. The soil specific fitting parameter β was 0.340, while the value of α was 0.325 m [Charbeneau et al., 1999, 2000]. Although this plot does not explicitly provide temporal extraction developments, they clearly demonstrate reductions in yield as H_o decreases over time.

LNAPL recovery as a function of time may be estimated by applying the principle of continuity and Equation (5.7) to a LNAPL body within the capture zone of a particular well. With the use of graphical nomographs or spreadsheets, estimates of recovery times can then be computed. These methods are straightforward to use but are usually based on numerous assumptions. An example of a simplified method was provided by Charbeneau et al. [1999, 2000] and is briefly explained here. A well with a capture radius r_c, extracting water and LNAPL, is located in the center of a LNAPL layer. The average LNAPL saturation, $\bar{S}_o$, is defined as:

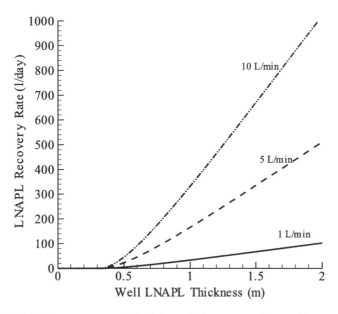

Figure 5.8. LNAPL recovery rate (L/day) for a single-pump well located in a sandy loam soil as a function of well LNAPL thickness (H_o) for three different water production rates. Parameter values are listed in the text.

$$\overline{S}_o = \frac{V_o}{\phi H_o} = \frac{\beta(H_o - \alpha)}{\phi H_o} \tag{5.8}$$

where Equation (4.23) has been used to substitute for V_o. Note that both V_o and H_o will reduce with time as LNAPL recovery proceeds. Then, the total volume of LNAPL, T_o, at any time is given by

$$T_o = \pi r_c^2 \phi(\overline{S}_o - (1 - \rho_{ro})S_{ru} - \rho_{ro}S_{rs})H_o \tag{5.9}$$

where S_{ru} and S_{rs} are residual saturations for the unsaturated and saturated regions, respectively. Substitution of (5.8) in (5.9) and rearrangement yields:

$$T_o = \pi r_c^2((\beta - \gamma)(H_o - \alpha\beta)) \tag{5.10}$$

where $\gamma = ((1 - \rho_{ro})S_{ru} + \rho_{ro}S_{rs})$. Next, the continuity equation for NAPL may be written as:

$$-\frac{dT_o}{dt} = Q_o \tag{5.11}$$

Substituting Equations (5.7) and (5.10) into the continuity equation yields:

$$\frac{dH_o}{dt} = -\left(\frac{\rho_{ro}\beta^2 Q_w}{\pi r_c^2 \mu_{ro}\phi^2 H_w(\beta-\gamma)}\right)\frac{(H_o-\alpha)^2}{H_o} \tag{5.12}$$

Now, let the initial value of H_o be denoted by H_o^i. Then, integration of Equation (5.12) yields:

$$\ln\left(\frac{H_o^i-\alpha}{H_o-\alpha}\right) + \frac{\alpha}{H_o^i-\alpha}\frac{H_o^i-H_o}{H_o-\alpha} = \left(\frac{\rho_{ro}\beta^2 Q_w}{\pi r_c^2 \mu_{ro}\phi^2 H_w(\beta-\gamma)}\right)t \tag{5.13}$$

The cumulative volume of extracted LNAPL at any time, U_o, can be obtained by subtracting T_o from its initial value T_o^i:

$$U_o = T_o^i - T_o = \pi r_c^2(\beta-\gamma)(H_o^i-H_o) \tag{5.14}$$

where Equation (5.10) has been employed. In principle, for a given situation with known fluids and properties, Equation (5.13), (5.14) and (5.7) can be used to calculate the variation with time of H_o, U_o, and Q_o, respectively. Equation (5.13) is, however, implicit in H_o. Therefore, it should be solved numerically.

Examples of how to use these equations in conjunction with a recovery nomograph are discussed in detail by Charbeneau et al. [1999, 2000]. They have also developed a computational procedure that can be easily implemented into a spreadsheet. Documentation about this procedure and the spreadsheet can be downloaded from the American Petroleum Institute website (http://groundwater. api.org/lnapl/). Although Charbeneau et al.'s [1999, 2000] procedure is widely used, it suffers from a number of limitations. In particular, it is sensitive to the time step size or the step size in the well LNAPL thickness. Here, we present an alternative procedure which is devoid of such limitations. A spreadsheet file (lnapl_recovery.xls) that can be used to calculate change with time in the LNAPL well thickness, LNAPL recovery rate, and extracted LNAPL volume is provided on the accompanying CD.

This computational procedure is easily implemented into a spreadsheet or a numerical program. However, users should realize that the outlined approach is based on several simplifying assumptions regarding the distribution of LNAPL and relative permeabilities, the homogeneity of the porous medium, constant residual saturations, and a linear relationship between the well LNAPL thickness and the specific LNAPL volume. For more complicated systems, multifluid flow models with well modules are recommended. Examples of simulators with the appropriate well routines are UTCHEM [Delshad et al. 1996] and STOMP [White and Oostrom, 2000].

To illustrate the procedure proposed here, extraction rates and cumulative extracted volumes produced by a single-pump well are computed for three different soils: sand ($\alpha = 0.10$ m; $\beta = 0.397$), loamy sand ($\alpha = 0.175$ m; $\beta = 0.363$) and sandy loam ($\alpha = 0.325$ m; $\beta = 0.340$). For these hypothetical examples, it was assumed that the porosity of each soil was 0.4, the radius of influence of the well 20 m, the saturated water thickness 20 m, the total liquid production rate 100 m^3/day, the residual LNAPL saturation 0.2, the relative LNAPL density 0.70, and the relative LNAPL viscosity 2.0. The initial well LNAPL thickness was 3.0 m. The results show (Figure 5.9 and Figure 5.10) that the reduction in the well LNAPL thickness is obtained the fastest in sand, followed by loamy sand and sandy loam. The pump in the sandy loam case has to be operational for almost twice as long as the sand case to reduce the well LNAPL thickness to 0.50 m. The total volume of extracted LNAPL (shown in Figure 5.11) also shows the same trend.

As is apparent from Equation (5.7), the computed LNAPL extraction rate, Q_o, for single-well pumps is very sensitive to the water extraction rate, Q_w. This sensitivity is shown in Figure 5.12 for a pumping scenario of LNAPL in a sand, where the water extraction rate ranges from 20 to 200 m^3. In this example, the

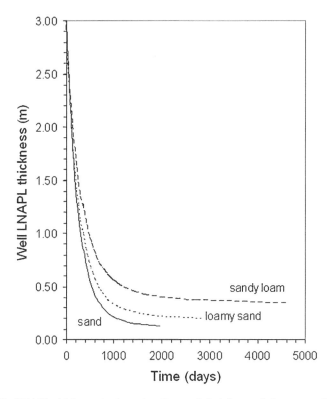

Figure 5.9. LNAPL thickness in the extraction well (m) for sand, loamy sand and sandy loam as a function of time.

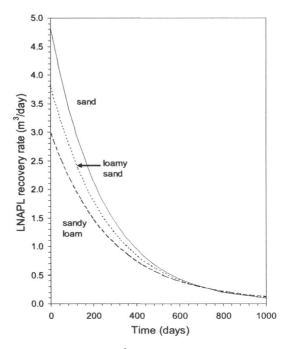

Figure 5.10. LNAPL extraction rate (m³/day) for sand, loamy sand and sandy loam as a function of time.

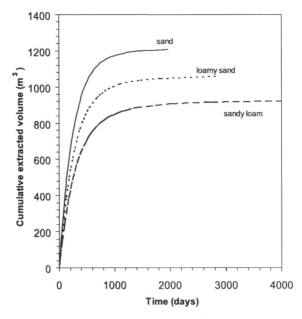

Figure 5.11. Cumulative LNAPL extraction volume (m³) for sand, loamy sand and sandy loam as a function of time.

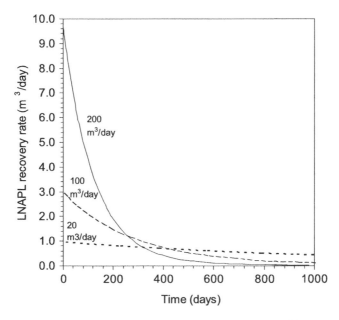

Figure 5.12. LNAPL extraction rate as a function of time for three different water pumping rate (20, 100, and 200 m³/day) for sand.

assumed LNAPL and sand properties are the same as defined previously in this section. Assuming identical pumps are used with the same settings, the differences in water extraction rates are the result of permeability variations of the sand. This example demonstrates that reasonable estimates of the water extraction rates should be available to compute meaningful predictions of removal times and LNAPL extraction rates.

5.2.4 Incomplete Removal of Residual LNAPL with Hydraulic Methods

Summary: Hydraulic methods, without chemical additives, are not effective in the removal of residual LNAPL. The lack of success using hydraulic recovery is demonstrated based on values of two non-dimensional numbers: the capillary number (N_{ca}) and the Bond number (N_B). Sometimes both numbers are combined into a total trapping number (N_T). For horizontal displacement, it is usually assumed that a capillary number larger than 10^{-4} - 10^{-5} is needed. For most hydraulic systems, the actual value might be orders of magnitude less than this critical value. When the displacement has a vertical component, the Bond number has to be evaluated too. For most LNAPLs in subsurface environments, the value of the Bond number is usually less than 10^{-6}. The relatively low value of the Bond number indicates that buoyancy forces by themselves do not result in residual saturation removal in a significant way.

Hydraulic recovery of NAPL relies on flow to a recovery well or trench under an induced hydraulic gradient. As NAPL is recovered, the remaining volume and saturations decrease, resulting in a lowering of the NAPL relative permeability. At some point in time, the NAPL saturation reaches residual values, at which the relative permeability becomes negligible, preventing further recovery. For a detailed discussion on residual saturation and related issues, the reader is referred to Chapter 2. The field residual saturation is considered to be the best-case end point of a recovery operation. It is obvious that recovery costs will increase when the field residual saturation is approached. In that case, the relative permeability of the remaining mobile NAPL becomes very small or considerable amounts of remaining NAPL might be located in zones not affected by the pumping scheme.

In an evaluation of hydraulic recovery cases, Huntley and Beckett [2002a] showed that the total NAPL recovery was less than 30% of the original volume released to the subsurface, with the upper end only as high as 60%. Some cases demonstrate even worse recovery efficiencies. For instance, they describe that in a 250-m^3 LNAPL spill in downtown San Diego, causing well thicknesses of up to 3 m, only 15 m^3 of NAPL have been recovered by aggressive pumping. Huntley and Beckett [2002a] also reported that at a former refinery in the central USA, only 13,200 m^3 out of a total spill of 24,000 m^3 have been removed from the subsurface after 13 years of pumping. The cost of recovery increased rapidly during the operation and went up from about $0.25 per liter initially to $12.50 in 2002. Future costs are expected to rise as a greater percentage of funds are allocated for water disposal per liter of LNAPL recovered.

During recovery operations, the fraction of the remaining LNAPL in residual form increases over time. Removal of the residual LNAPL by hydraulic methods is considered to be impractical. An understanding of the inefficiency of hydraulic methods in removing residual LNAPLs can be obtained by evaluating two applicable non-dimensional numbers, the capillary number, N_{ca}, and the Bond number, N_B.

The capillary number, N_{ca}, represents the ratio of viscous forces to the interfacial forces affecting the flow of fluid in porous media and is defined as:

$$N_{ca} = \frac{v_w \mu_w}{\sigma_{ow} \cos\theta} \qquad (5.15)$$

where v_w is the pore water velocity, μ_w the aqueous phase viscosity, σ_{ow} the LNAPL-water interfacial tension, and θ is the contact angle of the water-LNAPL interface. The number has been widely used in the petroleum industry to help in the design and operation of water-flooding and other enhanced oil recovery methods such as chemical floods with surfactants and alcohols.

Example relationships between N_{ca} and the displacement efficiency parameter E_d, in horizontal water-floods are shown in Figure 5.13. (adapted from Fig. 5.34 in Tiab and Donaldson [1996]). The parameter E_d, the ratio of the mobile to total LNAPL saturation, is defined as:

$$E_d = 1 - \frac{S_{or}}{1 - S_{wir}} \qquad (5.16)$$

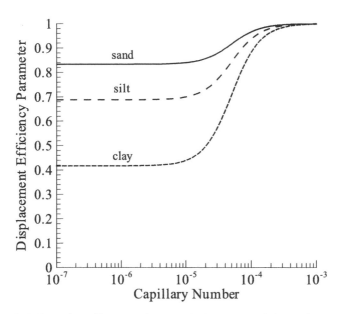

Figure 5.13. Effect of capillary number on displacement efficiency for a sand, silt and a clay.

where S_{or} is the residual NAPL saturation and S_{wir} is the irreducible water saturation. Figure 5.13 figure shows that below a capillary number of approximately 10^{-5}, the displacement efficiency does not change. Free NAPL can be replaced by infiltrating water with relative ease. However, the remaining NAPL is entrapped and discontinuous and therefore harder to replace. Only when the value of the capillary number becomes larger than 10^{-5}, a considerable increase of the efficiency and lower residual saturations are observed for all three porous media. Therefore, a critical value of 10^{-5} has been generally accepted for horizontal water displacement of NAPL in the petroleum industry [Tiab and Donaldson, 1996].

To illustrate what this critical value means for removal of residual NAPL in an aquifer, the relationship between N_{ca} and v_w is shown in Figure 5.14 for displacement of residual NAPLs with interfacial tensions σ_{ow} of 0.025 and 0.035 Nm^{-1}. Obviously the relationship between N_{ca} and v_w is linear (Equation (5.15)), and Figure 5.14 is shown to demonstrate the order of magnitude of N_{ca} values that can be expected for a range of v_w values. This range in interfacial tensions is typical for petroleum products. The figure shows that the critical value is reached for pore water velocities in the vicinity of 1000 m/day, which are beyond the realm of possibilities for normal recovery operations.

For displacements that are not horizontal, an additional non-dimensional number, the Bond number, N_B, also should be considered. The Bond number

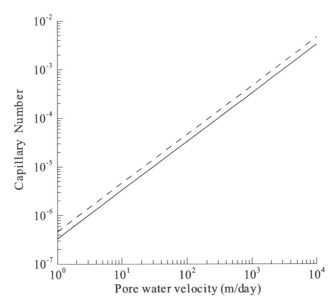

Figure 5.14. Relation between pore water velocity, v_w, and the capillary number, N_{ca}, for residual NAPLs wit.h NAPL-water interfacial tensions, σ_{ow}, of 0.035 Nm^{-1} (solid line) and 0.025 Nm^{-1} (dashed line) according to Equation (5.15). The aqueous viscosity is 10^{-3} Pa s.

is defined as the ratio of gravitational to capillary forces and is usually expressed as:

$$N_B = \frac{\Delta\rho gk}{\phi\sigma_{ow}\cos\theta} \qquad (5.17)$$

where $\Delta\rho$ is the density difference between water and the NAPL, g is the gravitational acceleration, k is the permeability, and n is the porosity. For illustrative purposes, a residual LNAPL with a density of 800 kg/m^3 and a σ_{ow} of 0.035 Nm^{-1}, located in a coarse-grained sand with a permeability of 10^{-10} m^2 and a porosity of 0.35, yields a N_B of 1.6×10^{-7}. This value indicates that for LNAPL recovery operations where the NAPL-water interfacial tension is not lowered, the influence of the N_B is limited. In some references, an alternative expression for N_B is used:

$$N_B = \frac{\Delta\rho gkk_{rw}}{\phi\sigma_{ow}\cos\theta} \qquad (5.18)$$

where k_{rw} is the water relative permeability.

A derivation that suggests the dependence of LNAPL saturation to these two non-dimensional numbers has been developed by Dawson and Roberts [1996]. They consider a non-deformable, rigid, homogeneous porous medium with two immiscible, incompressible fluids. For water and NAPL, the governing mass balance equations, without interphase mass transfer, are

$$\phi \frac{\partial S_i}{\partial t} = \nabla \cdot q_i \quad i = o, w \tag{5.19}$$

where ϕ is the porosity, S_i is the fluid saturation, q_i is the Darcy velocity, and t is time. The Darcy velocity is defined as

$$q_i = -\frac{k k_{ri}}{\mu_i} (\nabla p_i + \rho_i g \nabla z) \quad i = o, w \tag{5.20}$$

where p_i is the fluid pressure. Equation (5.19) is constrained by

$$S_o + S_w = 1 \tag{5.21}$$
$$p_c = p_o - p_w \tag{5.22}$$

where p_c is the capillary pressure. Differentiating Equation (5.22) yields

$$\nabla p_c = \nabla(p_o - p_w) \tag{5.23}$$

Rearranging Equation (5.20) and substituting into Equation (5.23) results in

$$\frac{1}{k}\left(\frac{q_w \mu_w}{k_{rw}} - \frac{q_o \mu_o}{k_{ro}}\right) - (\rho_o - \rho_w) g \sin \alpha = \nabla p_c \tag{5.24}$$

where α is the angle between the horizontal axis and the direction of flow. Next, we recast Equation (5.25) in a dimensionless form. Using the Leverett function we can define the dimensionless capillary pressure as

$$p_c^* = \frac{p_c}{\sigma_{ow} \cos \theta} \sqrt{\frac{k}{\phi}} \tag{5.25}$$

and introducing the dimensionless gradient

$$\nabla^* = \sqrt{\frac{k}{\phi}} \nabla \tag{5.26}$$

Equation (5.24) can be rewritten in non-dimensional form as follows

$$\frac{1}{k}\left(\frac{q_w \mu_w}{k_{rw}} - \frac{q_o \mu_o}{k_{ro}}\right) - (\rho_o - \rho_w) g \sin \alpha = \frac{\phi \sigma_{ow} \cos \theta}{k} \nabla^* p_c^* \tag{5.27}$$

Using $q_i/\phi = v_i$, Equation (5.27) can now be rewritten to yield

$$\frac{1}{k_{rw}} \frac{v_w \mu_w}{\sigma_{ow} \cos \theta} - \frac{1}{k_{ro}} \frac{v_o \mu_o}{\sigma_{ow} \cos \theta} - \frac{\Delta \rho g k \sin \alpha}{\phi \sigma_{ow} \cos \theta} = \nabla^* p_c^* \tag{5.28}$$

For immobile NAPL, $(v_o = 0)$ Equation (5.28) reduces to

$$\frac{1}{k_{rw}} \frac{\mathbf{v_w} \mu_w}{\sigma_{ow} \cos \theta} - \frac{\Delta \rho g k \sin \alpha}{\phi \sigma_{ow} \cos s\theta} = \nabla^* p_c^* \tag{5.29}$$

In terms of the non-dimensional N_{ca} and N_B numbers, Equation (5.29) can be written as

$$\frac{N_{ca}}{k_{rw}} - N_B \sin \alpha = \left| \nabla^* p_c^* \right| \tag{5.30}$$

In Equation (5.30), the right-hand side expresses the non-dimensional capillary pressure that has to be overcome in order to mobilize residual NAPL. Equation (5.30) suggests that LNAPL saturation, defined by the non-dimensionless capillary pressure term on the right-hand side of the equation, is a function of both the capillary number and the Bond number, provided the relative permeability of water in the presence of discontinuous NAPL is considered. The theoretical approach is consistent with proposed relationships by Morrow and Songkran [1981].

For horizontal displacement, Equation (5.30) is independent of the Bond number and becomes

$$\frac{N_{ca}}{k_{rw}} = \left| \nabla^* p_c^* \right| \tag{5.31}$$

Based on a derivation by Pennell et al. [1996], it can be shown that for vertical displacement of NAPL in the direction of the buoyancy force, N_{ca} and N_B can be combined into a total trapping number, N_T, by simple addition:

$$N_T = \left| N_{ca} + N_B \right| \tag{5.32}$$

Pennell et al. [1996] presented an empirical correlation relating the residual NAPL saturation to the total trapping number:

$$S_{or} = \frac{S_{or} |_{N_T=0}}{1 + (N_T / N_T^C)} \tag{5.33}$$

where $S_{or} |_{N_T=0}$ denotes the residual NAPL saturation at $N_T = 0$ and N_T^C is the critical trapping number.

Experimental work conducted by Pennell et al. [1996] yielded relationships between entrapped NAPL saturations and N_T similar to the graph shown in Figure 5.15 for two different porous media; a fine-grained and a medium-grained sand. The curves in the plot are constructed using Equation (5.33). For the sands and the DNAPL (perchloroethylene; PCE) used by Pennell et al. [1996], a critical N_T of about 5×10^{-4} was found, which is consistent with the value of 10^{-4} that is widely used as the critical N_{ca} for horizontal displacement of residual

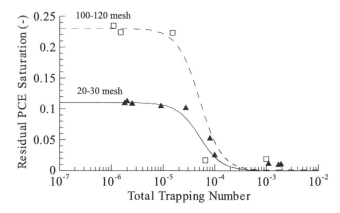

Figure 5.15. Comparison of experimentally determined residual PCE saturations and computed residual saturations using Equation (5.33) for a fine-grained (0.15-0.13 mm; circles) and medium-grained (0.85-0.6 mm; triangles) porous medium (after Pennell et al. [1996]).

LNAPL. In the work by Pennell et al. [1996], this critical value could only be reached or exceeded through addition of surfactants, resulting in a lowering of the interfacial tension between water and PCE. In order to obtain the critical value for N_T in their columns using clean water, the required pressure drop over the 13-cm long columns would have to be about 1 atm (10 m of water) to yield a flow rate of 3.8 m^3/day. These pressures and flow rates are not practical.

Besides affecting the mobilization of organic liquids, surfactant solutions may increase the effective solubility of organic compounds. The increased solubility is the main reason why surfactant remediation is sometimes considered as an alternative or enhancement of pump-and-treat remediation. Studies have shown that mobilization is a far more efficient recovery method than solubilization (see, e.g., Pennell et al. [1994]). However, utilization of this approach could lead to uncontrolled migration of the mobilized NAPL phase. The potential for displacement of NAPLs as a separate organic phase should be evaluated during the selection of surfactant formulations for aquifer remediation technologies. Such an evaluation should include an analysis of total trapping number values as a result of interfacial tension lowering and potential stabilizing effects of increased vertically upward flow rates.

5.3   PUMP AND TREAT

Summary: Extraction of contaminated groundwater and its treatment before being released is referred to as pump-and-treat. This process has been and is being used extensively as a tool to attempt groundwater remediation. Pump-and-treat has valuable capabilities and serious limitations. In this section, these capabilities and shortcomings will be discussed in the context of a case study.

A large number of sites with contaminated groundwater have been using groundwater extraction and treatment (pump-and-treat). Pump-and-treat systems have the objectives of controlling the movement of and removing contaminant mass from dissolved groundwater plumes. Both of these objectives are worthwhile; however the idea that pump-and-treat can be used as a cleanup technology has been largely discredited.

After pumping many millions of gallons of water over more than a decade, it has become clear that, at many sites, pump-and-treat has not been able to achieve the health-based cleanup goals that are often required for groundwater remediation. A few of these sites have been listed in Table 5.4. A more complete compilation of such sites has been reported elsewhere [National Research Council, 1994]. Some of the factors contributing to the inefficiency of the pump-and-treat are the inherent heterogeneity of subsurface media, the presence of low-solubility NAPLs, the penetration of contaminants into low-permeability material over a span of several years, and sorption and desorption of contaminants to and from geological materials. In some cases, the total volume of water extracted from the aquifer exceeds the original estimate of groundwater volume contaminated with dissolved NAPLs.

The following example illustrates an application of an enhanced pump-and-treat technique. The site, which has been contaminated with NAPLs, has been carefully characterized. Figure 5.16 shows a groundwater contamination plume in an area called "Old Town" at the Lawrence Berkeley National Laboratory in

TABLE 5.4. A short list of sites that have applied pump-and-treat for groundwater cleanup and have not achieved MCLs [EPA, 1998].

Site	Contaminants in Groundwater	Number of Years Pumped	Volume of Water Extracted $(m^3 \times 10^3)$
Des Moines TCE Superfund Site	TCE, DCE, VC	> 10	19,000
Former Firestone Facility Superfund Site*	PCE, TCE, DCE	> 14	> 6,800
JMT Facility	TCE, DCE, VC, TCA	> 10	> 190
Keefe Environ. Services	PCE, TCE, DCE	> 5	> 170
Lawrence Livermore Lab Site 300	TCE	> 7	> 350
Mystery Bridge Superfund Site	PCE, TCE, DCE, TCA	> 4	> 730
Old Mill Superfund Site	PCE, TCE, DCE, VC	> 9	> 50
SCRDI Dixiana Superfund Site	PCE, TCE, DCE, TCA	> 6	> 100

*USEPA Region 9 Website.

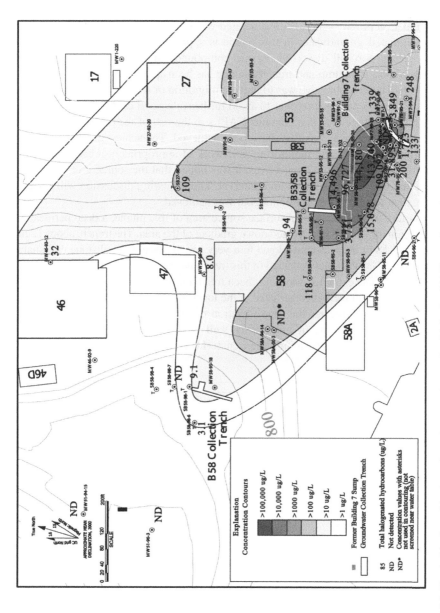

Figure 5.16. Total chlorinated hydrocarbons in groundwater (µg/L) in the Old Town plume (September 2002).

Berkeley, California. A color version of Figure 5.16 can be found on the accompanying CD in the file old_town_areal_plume.jpg. The principal contaminants originally released at this site during 1940s and 1950s were perchloroethylene (PCE), trichloroethylene (TCE) and carbon tetrachloride. Subsequently, other degradation products have been generated within the plume. The total volume of contaminated groundwater in this plume is approximately 50,000 cubic meters and the mass of all dissolved organic compounds in 1996 was approximately 7 kilograms. The pump-and-treat technique has been applied to this plume for two reasons. The first objective was to prevent migration of the plume to non-polluted areas. The second objective was to avoid further release of contaminants from the subsurface source area that contains NAPLs to the rest of the plume.

Both of the above objectives have been achieved. The achievements have been verified by periodically monitoring a network of appropriately located wells. In addition to the above two objectives, it was hoped that the source area would be depleted within a reasonable time interval. The following discussion will review the extent to which the additional goal was achieved.

In order to stabilize the subsurface source area of this plume, a deep permeable collection trench was constructed down gradient of the area. The trench, about 12 m long and 18 m deep, extends down to a very low permeability siltstone and shale formation. Groundwater collected in this trench has been extracted over a period of 45 months. The extracted water has been treated to non-detectable levels and then injected into a gravel pit constructed at the location of the original release site.

Figure 5.17 illustrates the time variation of total volatile organic compounds (VOCs) concentrations in monitoring well MW7B-95-21 located somewhere between the injection point and the trench. In September 2000, after more than

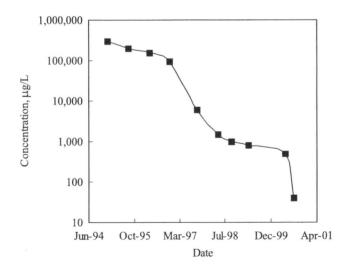

Figure 5.17. Time variation of total volatile organic compounds in MW7B-95-21.

3.5 years of pump-and-treat operation, concentrations of various VOCs in this well approached very close to the corresponding maximum contaminant levels (MCLs) for drinking water that are health-based benchmarks. At that time, to evaluate the adequacy of the remediation, the pump-and-treat program was continued but the recharge of the treated water was discontinued. Groundwater samples were collected from monitoring well MW7B-95-21 and tested for VOCs. Figure 5.18 shows the rebound of VOCs concentrations in this well: in less than 10 days VOCs concentrations in the monitoring well increased from 45 µg/L to about 15,000 µg/L. In about one month, total VOCs concentrations reached about 58,000 µg/L. The following discussion is an attempt to identify the causes of the above observation.

Figure 5.19 shows a geologic cross section through both the gravel pit and the collection trench. Four different geologic materials have been carefully mapped at this location. A color version of Figure 5.19 can be found on the accompanying CD in the file old_town_ cross_section.jpg. Close to the ground surface, there is an artificial fill overlaying Moraga Formation: an intensely fractured volcanic rock stratum. The material under the volcanic rock is a mixture of volcanic materials and the sedimentary rocks that underlie it. The fractured volcanic rock has a relatively high hydraulic conductivity of the order 10^{-5} m/s, while the mixed unit under that has a much lower conductivity of the order of 10^{-7} m/s. The sedimentary rock beneath the mixed unit has a very low conductivity of the order 10^{-9} m/s. Both the volcanic and the mixed units are intercepted by the gravel pit and thus received the recharged water. Groundwater velocity in the volcanic rock unit is about 100 times faster than that of the mixed unit.

The approximate travel time between the gravel pit and the trench, within the volcanic unit, is estimated to be about 50 days. Therefore, in 45 weeks more than six pore volumes of water traveled through the volcanic rock units and carried dissolved phase and most of the adsorbed VOCs to the trench. Meanwhile, most

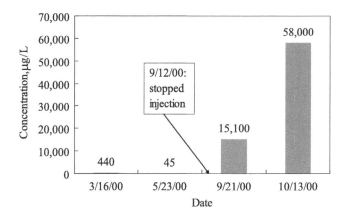

Figure 5.18. Time variation of total VOCs before and after stopping injection of clean water.

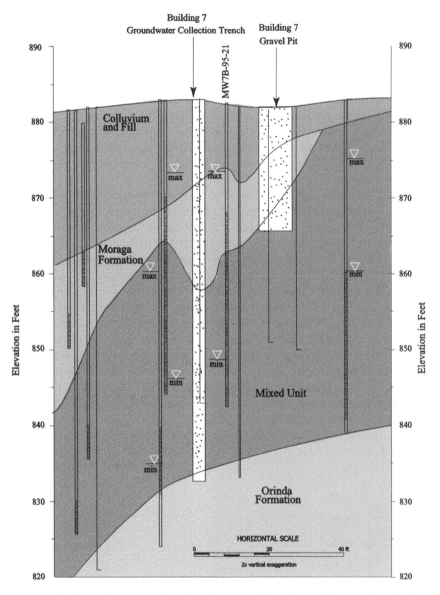

Figure 5.19. Geologic cross section through the Building 7 gravel pit and the collection trench.

of the initial contaminants still remained in the mixed unit. Monitoring well MW7B-95-21 is screened in both volcanic and mixed units. When recharge into the pit was in progress, groundwater flowed in both volcanic and mixed units. At that time, approximately 99% of the water entering the monitoring well arrived from the volcanic unit and about 1% came from the mixed unit. At the end of 45 weeks of operation, the VOCs concentration in the groundwater passing through the volcanic unit had dropped to a very low level, and the high VOCs concentration water entering the well from the mixed unit was diluted by about one to ninety-nine. Once the recharge of clean water into the pit was discontinued, the flow of water through the volcanic unit was diminished and an increasingly higher percentage of water entering the monitoring well came from the mixed unit that was still highly contaminated.

Soil water samples collected from the mixed unit at the end of the above operation revealed total VOCs concentrations up to 380,000 μg/L. Considering that the solubility of PCE, the main constituent present in this area, is about 150,000 μg/L, chances are that at least a small amount of PCE in NAPL form is present. Let us now review how contaminants could penetrate such a low-permeability material and what the role of these low-permeability materials is in the efficiency of pump-and-treat as an aquifer cleanup method.

Contaminant transport into such low-permeability materials can occur through both advection and diffusion. Generally, for high-permeability materials, the role of diffusion relative to advection is negligible. For very low permeability materials, however, diffusion may become a significant mode of transport. For many years, NAPLs were routinely released in this area. The presence of NAPLs and/or groundwater with very high VOCs concentration in the volcanic unit had established a very high concentration gradient between the volcanic and the mixed units. Therefore, both advection and diffusion helped to transport contaminants into the mixed zone.

Let us consider a two-layer flow system as shown in Figure 5.20. Due to the major contrast of hydraulic conductivity between the two layers, equipotentials are essentially vertical in the top layer and horizontal in the lower layer. That

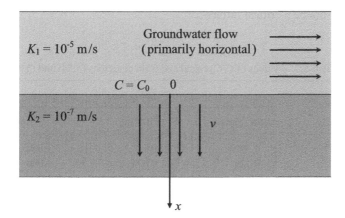

Figure 5.20. Schematic diagram of a two-layer flow system.

means for all practical purposes, flow in the top layer is horizontal and flow in the lower layer is vertical. Now, let us consider a simple one-dimensional advection-diffusion transport model from the top layer into the lower one. If C_0 represents the concentration of a solute in the top layer, the concentration distribution of the solute, $C(x, t)$, in the mixed unit may be represented by

$$\frac{C}{C_0} = \frac{1}{2}\text{erfc}\left(\frac{x - vt}{2\sqrt{Dt}}\right) + \frac{1}{2}\exp\left(\frac{vx}{D}\right)\text{erfc}\left(\frac{x + vt}{2\sqrt{Dt}}\right) \qquad (5.34)$$

as given by Ogata [1970], where x is the vertical distance into the mixed unit, t is time, D is the diffusion coefficient, v is the pore water velocity in the mixed unit, and erfc is the complementary error function. If we assume that the pore water velocity is negligible and as a result molecular diffusion is the sole mode of contaminant transport into the mixed unit, Equation (5.34) reduces to

$$\frac{C}{C_0} = \text{erfc}\left(\frac{x}{2\sqrt{Dt}}\right) \qquad (5.35)$$

Now assuming that concentration of the solute at the contact of the volcanic and mixed units is given by C_0, Table 5.5 presents values of C/C_0 calculated from Equations (5.34) and (5.35) for $t = 50$ years, $D = 5 \times 10^{-10}$ m^2/s, and 6 values of x ranging between 0.5 and 3 meters and 3 values of pore water velocities in the low-permeability layer. Table 5.5 reveals that although molecular diffusion is a very slow process, after 50 years, contaminants may penetrate significant distances into the mixed unit. A spreadsheet file (twolayer.xls) that can be used to make calculations of concentrations in a low permeability layer adjacent to a high permeability layer with constant concentration is found on the accompanying CD. Note that in the above calculation the adsorption effect was not taken into account. This may not introduce significant error because, generally speaking, adsorption of chlorinated hydrocarbons, when total organic carbon in the rock is small, may not be a critical influence.

TABLE 5.5.   Values of C/C_0 calculated from equations (5.34) and (5.35).

x (m)	C/C_0		
	$v = 0.0$ m/s	$v = 10^{-9}$m/s	$v = 10^{-8}$ m/s
0.5	0.689	0.937	1.000
1.0	0.424	0.824	1.000
1.5	0.231	0.665	1.000
2.0	0.110	0.485	1.000
2.5	0.046	0.315	1.000
3.0	0.016	0.180	1.000

Although the above process may lead to relatively high values of VOCs concentrations after a period of 50 years, it does not explain concentrations above the solubility of PCE inside the mixed unit, as reported earlier. The reason for such high concentration in the mixed unit may be explained by the presence of some relatively thin but highly permeable zones inside the mixed unit. Unfortunately, our ability to discover the presence of such features within the body of low-permeability masses is limited. The most common in-situ hydraulic tests that one can perform in low-permeability media are the single-well tests. The zone of influence of these tests is very small and they are not capable of identifying relatively thin but high permeability layers that are present at small distances from the test area. In fact, the data obtained from a soil heating experiment conducted recently in the mixed unit confirmed, without any doubt, that some discrete high permeability features exist in this unit that can easily conduct fluids within certain part of this rock mass.

Now, let us go back and attempt to find an answer for the inefficiency of the pump-and-treat technique for aquifer cleanup. Very often unconsolidated aquifers consisting of sand and gravel contain lenses of silty clay or clayey silt of various thicknesses. Even in rare cases where aquifers themselves do not contain clay lenses, very low permeability layers often constitute the lower part of the aquifers. Under either of these two scenarios, when NAPL releases occur over a long period of time, at least some of the dissolved constituents penetrate these low permeability zones. During the remediation phase, once several pore volumes of water pass through the contaminated section of the aquifer and advection and desorption processes remove most of the contaminants, the solute concentration in the aquifer decreases. This will generate a concentration gradient from the lower permeability zone toward the aquifer that results in "back diffusion" of contaminants from lower permeability units to the aquifer. In this case, these previously contaminated low permeability zones act as long-term reservoirs for polluting aquifers. The magnitude of the impact of this process on the quality of groundwater down gradient from such areas depends on the groundwater velocity in the aquifer and the rate at which contaminant mass is being released to the aquifer.

During the "back diffusion," because of the lower gradient, the rate of contaminant release is smaller than the entry rate into the low-permeability materials. Therefore, if it took 50 years for contaminants to enter the mixed zone, under hydraulic processes only, it would take much more than 50 years to get rid of all contaminants in the system.

An important assumption in the above discussion was that no more NAPLs are present either in the aquifer or within the bulk of low permeability zones. In cases where the original NAPL still is present in the system, the required remediation time will be much longer. Since the aqueous solubilities of NAPLs are generally very low, depending on the groundwater velocity and the volume of available NAPLs, it may take years or even decades to remove such materials through dissolution into the groundwater and subsequent extraction of groundwater.

In conclusion, the case shows that the pump-and-treat technique, if properly designed, could play an important role in preventing the groundwater contamination plume from extending into non-contaminated areas. It can also contain the source area if it is correctly characterized. If one applies the pump-and-treat technique extensively over the tail end of the plume, away from the source area, it could also help to shrink the size of the plume. This has been demonstrated at the Lawrence Livermore National Laboratory site in Livermore, California over a period of more than 10 years of pump-and-treat operation [Berg et al., 2002]. For VOCs-contaminated groundwater plumes, pump-and-treat may not be an efficient technique for cleanup of source areas that contain NAPLs and/or high concentrations of contaminants.

5.4   SOIL VAPOR EXTRACTION

Summary: Gas concentrations measured at SVE wells are almost always lower than equilibrium values predicted from soil concentrations. This relative inefficiency is due to a combination of gas dilution, local scale kinetic interphase mass transfer limitations, and larger scale gas diffusion-limited mass transfer.

Soil vapor extraction (SVE) is an effective and well established remediation technique for the removal of volatile organic compounds from the vadose zone [see, e.g., Looney and Falta, 2000]. An SVE system consists of one or more wells screened above the water table, connected to a blower. A vacuum is applied to the wells, inducing gas flow through the vadose zone. This gas flow results in evaporation and volatilization of contaminants, with gas phase transport towards the extraction well. In many situations, SVE wells have removed thousands of kilograms of contaminants from the subsurface over a period of a few months. In other cases, however, low contaminant concentrations in the SVE well effluent have been encountered, which means that the efficiency of SVE operations has been much lower than expected. Furthermore, in virtually all field applications, the concentration of vapors in the SVE well effluent has been observed to decline with time, often exponentially. The observed effluent gas concentrations in SVE wells are always below the theoretical equilibrium values for the contaminant, often by one order of magnitude or more.

A good example of this concentration decline with time is provided by Hiller [1991], who summarized the performance of six full-scale SVE applications involving TCE and PCE. These compounds are volatile, with vapor pressures of about 7800 Pa (TCE) and 2500 Pa (PCE) at room temperature. These vapor pressures correspond to volume fractions of about 78,000 parts per million by volume (ppmv) for TCE and 25,000 ppmv for PCE. The initial SVE effluent concentrations of PCE and TCE in Hiller's review ranged from about 100 ppmv to about 800 ppmv. Therefore, in these real field applications, the initial SVE concentrations were on the order of one percent or so of the equilibrium values if free NAPL were present. In each case, the concentrations dropped rapidly with time over the first 20 days of venting. The decline in concentrations during this period ranged from about 50% up to more than 90%. After 20 days, the concentrations tended to stabilize somewhat, but continued to drop for several more months.

Several mechanisms have been proposed to explain the relatively low vapor concentrations observed in SVE operations. These mechanisms include micro-scale mass transfer limitations due to aqueous phase diffusion and kinetic desorption, macro-scale mass transfer limitations due to gas phase diffusion in heterogeneous formations, effects of limited contaminant extent and the contaminant geometry, and dilution due to mixing with atmospheric air. While these theories have been studied extensively in laboratory and through theoretical investigations, it is rarely possible to conclusively isolate the exact reason for the relatively low observed vapor concentrations in a real field SVE application. This is partly due to the fact that all of the above mechanisms tend to result in SVE effluent gas concentrations that are well below equilibrium values, and they all tend to lead to an exponential decline with time.

A starting point for the understanding of SVE vapor concentrations is the calculation of theoretical maximum equilibrium values for a particular site. This topic was covered in some detail in Chapter 2, and is reviewed here with attention focused on gas phase equilibrium concentrations.

For sites where a single-component NAPL is known to be present, the maximum gas concentration is given by the ideal gas law (equation (2.36)) using the chemical's pure vapor pressure at the prevailing subsurface temperature:

$$\overline{C}_g^i = \frac{P_{vap}^i M_{wt}^i}{RT} \tag{5.36}$$

The mass concentration given by Equation (5.36) may be converted to other measures of gas concentration (volume fraction, mole fraction, partial pressure) using the methods described in Chapter 2.

If a multicomponent NAPL is known to be present at a site, the maximum gas concentration is calculated with Raoult's Law (combination of Equations (2.36) and (2.38)):

$$\overline{C}_g^i = \frac{\chi_n^i P_{vap}^i M_{wt}^i}{RT} \tag{5.37}$$

where χ_n^i is the mole fraction of the component of interest in the NAPL phase. In the event that subsurface contamination levels are below those indicative of NAPL presence, the maximum gas concentrations may be estimated by consideration of multiphase equilibrium, using the total concentration concepts described in Chapter 2. When NAPL is not present, and if linear equilibrium partitioning is assumed, then the gas concentration is a linear function of the soil total concentration as shown previously in Equation (2.54). If the aqueous concentration is known, then the equilibrium gas concentration is calculated from Henry's law, Equation (2.39).

An important implication of Equation (5.37) is that non-volatile components, or components with low vapor pressures, will be found in the gas phase only at low concentrations, even if the mole fraction of the component is relatively high. The low concentration in the gas phase for non-volatile components means that these components will be removed only sparingly by SVE. For example, the vapor pressures of benzene and n-octadecane are 0.13 and 2.1×10^{-7} atm, respectively, at 25°C. If all other parameters (except molecular weight) are the same, we find from Equation (5.37) that the gas phase concentration of benzene is more than five orders of magnitude higher than that of n-octadecane.

Because SVE operates with a moving gas phase, the limited contact time between the flowing gas and contaminants can lead to conditions where the local concentration in the gas phase is less than the equilibrium concentration. Here, the term "local" refers to a scale of a few millimeters or centimeters, and this phenomenon is known as kinetic interphase mass transfer. Under the nonequilibrium conditions of kinetic interphase mass transfer, the rate of mass transfer is limited compared to situations where phase concentrations are in equilibrium. The topic of kinetic mass transfer during sparging has been the subject of a number of experimental studies [Baehr and Hoag, 1988; Gierke et al. 1992; Wilkins et al., 1995; Fischer et al., 1996].

Mathematically, the process of local kinetic interphase mass transfer is often represented as a first order heterogeneous reaction [Gierke et al., 1992; Wilkins et al., 1995]:

$$Q^i_{mt} = k^i_{mt}a(\bar{C}^i_g - C^i_g)$$ (5.38)

where Q^i_{mt} is the rate of chemical mass transfer into the gas phase per unit volume of porous media, $k^i_{mt}a$ is the mass transfer coefficient-interfacial area product (also known as a lumped mass transfer coefficient) defined in Section 4.2.1, and $\bar{C}^i_g$ is the equivalent equilibrium gas concentration. In Equation (5.38), the rate of mass transfer is a linear function of the mass transfer coefficient, and of the degree of nonequilibrium. As the lumped mass transfer coefficient approaches 0.1 s^{-1} or more, the mass transfer process is essentially an equilibrium process. Small values of the mass transfer coefficient, below roughly 10^{-4} s^{-1}, indicate more severe mass transfer limitations [Fischer et al., 1996]. Unfortunately, the mass transfer coefficient for a particular system is generally unknown, and it depends on the local details and scale of the mass transfer regime.

Considering a simplified scenario of 1-D flow, and neglecting dispersion, diffusion, and transient effects, a differential equation for the gas concentration as a function of contact distance can be written as (see, e.g., Hunt et al. [1988])

$$v_g \frac{dC^i_g}{dx} = k^i_{mt}a(\bar{C}^i_g - C^i_g)$$ (5.39)

If clean air is entering the system at $x = 0$, then the concentration as a function of contact distance is [Hunt et al., 1988]

$$\frac{C_g^i}{\bar{C}_g^i} = \left\{ 1 - \exp\left(\frac{-k_{imt}a}{v_g} x \right) \right\} \tag{5.40}$$

Equation (5.40) can be rearranged to calculate the distance required for a given amount of mass transfer. Table 5.6 shows the contact distance required to obtain different gas concentrations as a function of the mass transfer coefficient. These values were obtained with a Darcy velocity of 2 m/day, a porosity of 0.3, and a gas saturation of 0.67. This represents a reasonable upper range for the gas velocity under field SVE operations (see, e.g., Falta et al. [1993]). A spreadsheet file (contact_mt.xls) that can be used to make calculations of distance required to achieve a given concentration level in the gas phase is found on the accompanying CD.

The observed rate of mass transfer in SVE laboratory experiments varies over a wide range. In an experiment in which gasoline was evaporated by SVE, Baehr et al. [1989] found that the mass transfer to the gas phase was essentially an equilibrium process. Similarly, Wilkins et al. [1995] studied the evaporation of styrene and other chemicals at pore air velocities ranging up to 1.5 cm/sec or 1296 m/day. At these very high velocities, they found that the effluent concentration leaving a 10-cm-long column, was only at about 60% to 80% of the equilibrium value. However, at more realistic pore velocities of 10 m/day or so, their data show effluent concentrations on the order of 90-100% of equilibrium values. Referring to Table 5.6, these two sets of experiments suggest that the local scale interphase mass transfer coefficient, $k_{mt}^i a$, for NAPL evaporation may be on the order of 10^{-2} to 10^{-3} s^{-1}.

In cases where free NAPL is not present, diffusion through a local liquid water film may reduce the rate of mass transfer under some conditions. In a study of toluene vapor transport in variably saturated columns, Gierke et al. [1992] found chemical removal from moist sands was not affected by nonequilibrium. However, this same study found that nonequilibrium effects were significant for vapor transport in aggregated porous soils, where a $k_{mt}^i a$ value of 6.8×10^{-3} s^{-1} gave a best fit of the data.

TABLE 5.6.   Contact distance in meters required for different levels of mass transfer.

$C_g^i / \bar{C}_g^i$	$k_{mt}^i a, s^{-1}$			
	1 s^{-1}	10^{-2} s^{-1}	10^{-4} s^{-1}	10^{-6} s^{-1}
0.99	5.30E − 04	5.30E − 02	5.30E + 00	5.30E + 02
0.9	2.65E − 04	2.65E − 02	2.65E + 00	2.65E + 02
0.5	7.98E − 05	7.98E − 03	7.98E − 01	7.98E + 01
0.1	1.21E − 05	1.21E − 03	1.21E − 01	1.21E + 01

Fischer et al. [1996] studied SVE removal of dissolved chlorinated solvents from 2-D laboratory sand tanks. Their simulations of the experiments indicated that the apparent local mass transfer coefficient changes during the course of the SVE operation. During the early time, when the gas concentration in sampling ports ranged from $\bar{C}_g^i$ down to about 0.2 $\bar{C}_g^i$, an equilibrium modeling approach matched the concentration versus time data. Later, when the concentrations dropped to about 0.05 $\bar{C}_g^i$, then, a $k_{mt}^i a$ value of about 10^{-4} s^{-1} was needed to model the experiment. In some of the experiments, $k_{mt}^i a$ values as low as 10^{-6} s^{-1} were needed to match the very low concentration tails of the concentration versus time curves at large times. They attributed the apparent decrease in $k_{mt}^i a$ with time to the increase in the average liquid diffusion length with time.

The results of various studies on local nonequilibrium during SVE suggest that it is probably not important to consider it in cases where NAPL is present. Nonequilibrium effects may be significant in cases where the dissolved contaminant concentrations are low, or at large times. It is important to recognize, however, that local scale mass transfer effects that are observed at the small laboratory scale may not be dominant at the field scale where macro-scale flow and mass transfer become important.

A fairly obvious reason for SVE effluent concentrations to be below equilibrium values is the subsurface and in-well mixing of the vapors with clean air. In homogeneous systems, this vapor dilution can occur if the well "capture zone" is not fully contaminated. It can also occur due to short-circuiting of clean air from the ground surface to the well through a zone in which the contaminant has already been removed.

The effect of the initial contaminant geometry on the SVE effluent curve can be illustrated by a simple example. Considering radial flow of a compressible gas to a fully screened well in an unsaturated system confined both above and below, Falta et al. [1993] calculated the gas stream function and trichloroethylene (TCE) travel time distributions. The stream function can be used to define streamtubes through which a certain fraction of the gas flow occurs. These are shown in Figure 5.21, and they correspond to a case where the well is maintained at an absolute pressure of 0.7 atm in the formation with an intrinsic permeability of 10^{-12} m^2, and a gas saturation of 0.75. The horizontal lines in this figure are the contours of the normalized stream function. In this simple example, 10% of the gas flow occurs between each stream function contour. The vertical lines are contours of the TCE travel time to the well in units of days (including the effects of phase partitioning with a gas phase retardation factor of 3.73). Thus, the TCE vapor located within a radius of about 6m would reach the well in one day, and TCE vapor located at a radius of about 11 m would take about 3 days to arrive at the well.

Figure 5.21 also shows an idealized TCE contaminated zone geometry. By mathematically transforming the contaminated zone from the r-z plane to the stream function-travel time plane, it is possible to directly calculate the theoretical breakthrough curve, which is shown in Figure 5.22. Here, the SVE effluent

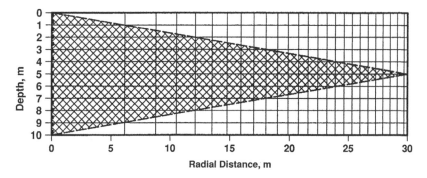

Figure 5.21. Normalized streamfunction and gas vapor travel time for radial flow case (from Falta et al. [1993]).

concentration drops with time as the contaminated vapors move by advection towards the well, and are mixed with clean air.

A more realistic flow geometry is shown in Figure 5.23, which gives the theoretical normalized stream function distribution for the case of an SVE well located above the water table, with an open ground surface [Shan et al., 1992]. In this figure, each of the streamline contours represents 5% of the gas flow to the well, and all of the gas originates at the atmosphere, where it is presumably clean. It is also possible to numerically compute the gas travel time from any location in this figure to the well screen. Figure 5.24 shows the unretarded gas travel time from the ground surface to the well for each of the streamlines shown in Figure 5.23 assuming an SVE well flow rate of 0.1 kg/s (0.08 m^3/s or 171 cfm), a permeability of 1×10^{-11} m^2 and a volumetric gas content of 0.4. It also gives travel times for an anisotropic case where the vertical permeability has been reduced by a factor of 10.

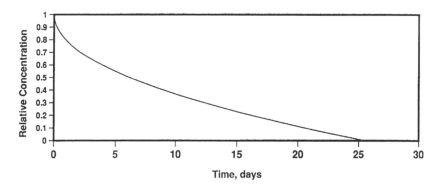

Figure 5.22. SVE effluent concentration for the radial flow case (from Falta et al. [1993]).

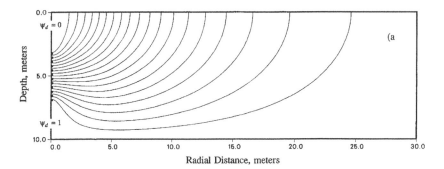

Figure 5.23. Normalized stream function distribution for gas pumping in a system open to the atmosphere (from Shan et al. [1992]).

For the conditions of this problem, in the isotropic case, atmospheric air reaches the well screen in a very short time, about 1000 seconds. After 10,000 seconds, 40% of the air reaching the well screen has originated at the ground surface, and after a little over a day, almost 80% of the air reaching the well screen has originated at the ground surface. While contaminants will not move as fast as the gas due to multiphase retardation, it is clear that regions above the well screen will quickly be cleaned, leading to dilution from the atmospheric air. An SVE effluent concentration versus time curve can be produced for the case of uniform subsurface contamination by subtracting the curve in Figure 5.24 from one. The effects of chemical retardation can be included by multiplying the time scale by the gas phase retardation factor. This curve would show an exponential decrease in the SVE well concentration over time simply due to dilution.

Figure 5.25 [Falta et al., 1993] shows the contours of unretarded gas travel time to the well screen for a similar case, but with an increased SVE extraction rate of 0.155 m^3/s (328 cfm) and a smaller volumetric gas content of 0.3. Here, the travel time contours correspond to a travel time of one day. As in the previous radial flow example, it is possible to mathematically transform any initial contaminant geometry from the r-z plane to the stream function-travel time plane. Figure 5.26 shows the theoretical SVE effluent concentration as a function of time for the contaminated zone geometry shown in Figure 5.25. The effects of dilution and of the initial contaminant geometry are evident.

Gas flow during SVE operations is strongly affected by heterogeneities in the form of permeability and capillary pressure variability. Fine-grained sediments such as clays are characterized by both low intrinsic permeability and high capillary pressure, leading to higher natural water contents and lower gas phase relative permeabilities. In contrast, coarse-grained sediments such as sands have higher intrinsic permeabilities and lower capillary pressures, giving them much higher effective gas phase permeabilities. For these reasons, it is expected that the SVE-induced gas flow will occur primarily in the coarse grained media, and that

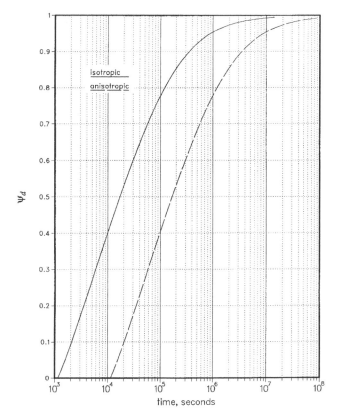

Figure 5.24. Gas travel time from the ground surface (from Shan et al. [1992]) for various streamlines shown in Figure 5.26 for cases of isotropic and anisotropic porous media.

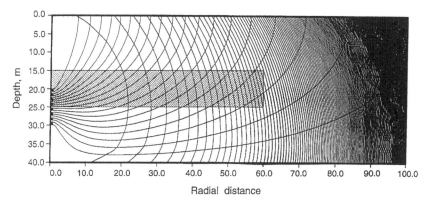

Figure 5.25. Gas travel time contours for example with an open ground surface (from Falta et al. [1993]).

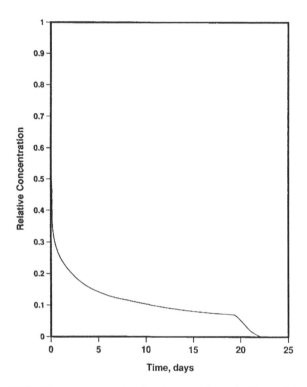

Figure 5.26. SVE effluent concentration for the case shown in Figure 5.25 [Falta et al., 1993].

low permeability, high capillary pressure fine-grained media will be bypassed by the flow (see Figure 5.27).

If contaminants are primarily found in the coarse materials, they can be efficiently removed by SVE operations. This could likely be the case at the site of a recent spill, where the contaminants have not penetrated the fine-grained sediments. However, at older sites, much of the remaining contamination is found in low permeability layers and lenses that are not effectively swept by the flowing gas. Under these conditions, experimental studies by Ho and Udell [1991, 1992], and theoretical analyses by Johnson et al. [1990a, 1990b] and Ho and Udell [1991] have shown that the mass transfer from the low permeability layers into the flowing gas can be rate limited by macro-scale gaseous diffusion.

Figure 5.28 shows a diagram of the mass transfer regime considered in the Ho and Udell [1991] analytical solution. They consider two regions: a high-permeability zone in which the gas flows, and a low-permeability zone with no advective gas flow. The contaminant, which may exist at soil concentrations above or below NAPL forming concentrations, is only located in the low-permeability zone, and at time zero, it is present at the interface with the high-permeability

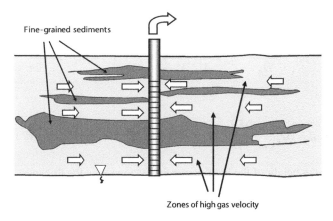

Figure 5.27. Schematic illustration of gas flow patterns in heterogeneous porous media.

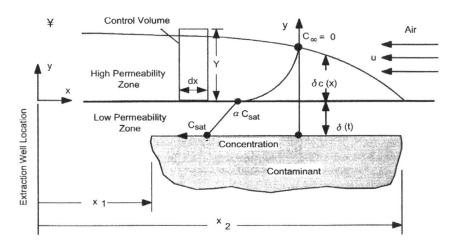

Figure 5.28. Diagram of the concentration boundary layer considered in the diffusion limited analytical solution by Ho and Udell [1991].

zone. With time, the contaminant interface retreats into the low-permeability zone and, thus, diffusive resistance in both the low- and high-permeability zones has to be taken into account.

The Ho and Udell [1991] solution for a radial gas flow system is written as:

$$\frac{C_g^i}{\bar{C}_g^i} = \frac{9}{8}\frac{F}{Sh^*}\frac{1}{\sqrt{t^*}}$$

with

$$Sh^* = \frac{\overline{h}_m H}{D_p}; \quad \overline{h}_m = \frac{3}{4}\left(\frac{2QD_p}{V_a}\right)^{1/2}; \quad t^* = 1 + \beta t; \quad F = \frac{\theta}{2\pi} \qquad (5.41)$$

and

$$\beta = 2\overline{C}_g^i\left(\frac{V_s}{m_o}\right)\frac{\overline{h}_m^2}{D_i}; \quad V_a = H\pi(R_2^2 - R_1^2); \quad V_s = bF\pi(R_2^2 - R_1^2)$$

where H is the height of the high-permeability zone, b is the thickness of the contaminated zone in the low-permeability layer, Q is the volumetric flow rate of gas to the SVE well, R_1 is the radial distance to the start of the contaminated zone, R_2 is the radial distance to the end of the contaminated zone, m_o is the initial mass of contaminant in the low-permeability zone, D_p is the effective gas diffusion coefficient in the high-permeability zone, D_i is the effective gas diffusion coefficient in the low-permeability zone, and θ is the horizontal angle subtended by the contaminated zone (equal to 2π if the contamination surrounds the extraction well).

Figure 5.29 shows the results of Equation (5.41) for a case where the gas flow rate is 0.0236 m^3/s (50 cfm), the high-permeability zone is 1 m thick, the

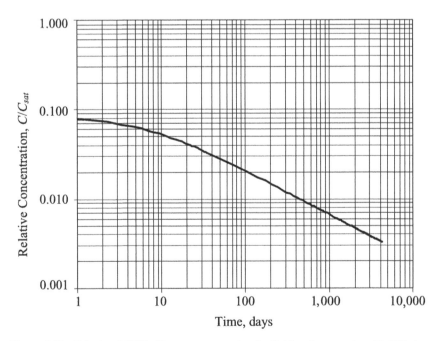

Figure 5.29. Calculated SVE effluent concentration for field scale example with diffusion limited mass transfer.

contaminated zone is 2 m thick, with a NAPL saturation of 0.05, a porosity of 0.35, a NAPL saturated vapor concentration of 0.0391 kg/m^3, a NAPL density of 800 kg/m^3, and an initial contaminant mass of 2200 kg. The effective gas diffusion coefficients in both materials were set at 1.95×10^{-6} m^2/s, and the inner and outer radii of contamination were 0 and 5 meters, respectively. The results in Figure 5.29 show that, even at early times, the expected SVE effluent concentration is only about 8% of the equilibrium value. Over a period of 4000 days, this concentration drops to only about 0.3% of the equilibrium value, even though some NAPL is still present in the low permeability zone. A spreadsheet file (houdell.xls) that can be used to make calculations of concentrations in a soil vapor extraction well for a two-layer system is found on the accompanying CD.

In most cases, SVE will be most effective if the media is relatively homogeneous and permeable. In this case, the gas flow can be controlled by the system designer to maximize flow through all contaminated zones. This becomes substantially more difficult in highly heterogeneous systems if a large fraction of the contamination is present in low permeability zones. In this case, little can be done to improve the flow of gas through these regions.

Notation

Latin Symbols

A	cross-sectional area, $[L^2]$
b_c	thickness of the contaminated zone in the low permeability layer, $[L]$
b_w	aquifer thickness, $[L]$
C_β^i	mass of chemical i per unit volume of phase β ($\beta = a, o, s, w$), $[ML^{-3}]$
C_T^i	total mass concentration, $[ML^{-3}]$
$\bar{C}_T^i$	maximum possible value for C_T^i in the absence of NAPL, $[ML^{-3}]$
$\bar{C}_g^i$	equivalent equilibrium gas concentration, $[ML^{-3}]$
$\bar{C}_w^i$	aqueous solubility of a pure compound i, $[ML^{-3}]$
c_β	phase molar density, $[mol\,L^{-3}]$
c_β^i	molar concentration: moles of component i per unit volume of phase β, $[mol\,L^{-3}]$
D	DNAPL pool depth, $[L]$
D_{min}	minimum DNAPL pool depth for entry into finer porous media, $[L]$
D_o	soil hydrocarbon thickness, $[L]$
$D_{w,l}$	longitudinal dispersion coefficient, $[L^2T^{-1}]$
$D_{w,t}$	transverse dispersion coefficient, $[L^2T^{-1}]$
D_β^i	chemical diffusion coefficient in free fluid phase β ($\beta = a, o, w$), $[L^2T^{-1}]$
$D_a^{i\,eff}$	effective diffusivity of chemical component i in the gas phase, $[L^2T^{-1}]$
D	distance between the source point and the destination point or a specified minimum distance, $[L]$
d	mean grain size, $[L]$
E	exponent associated with inverse distance method,
E_d	displacement efficiency parameter
e	fracture aperture, $[L]$
$\bar{F}_o$	forces acting on NAPL phase, $[MLT^{-2}]$
$\bar{F}_w$	forces acting on water phase, $[MLT^{-2}]$

$\mathbf{F}_\beta^i$	advective mass flux for chemical component i in phase β ($\beta = a, o, w$), $[ML^{-2}T^{-1}]$
g	gravitational acceleration, $[LT^{-2}]$
$\mathbf{g}$	gravitational vector, $[LT^{-2}]$
f_{or}	LNAPL reduction factor
H	height, $[L]$
H_o	well hydrocarbon thickness, $[L]$
H^i	Henry's coefficient, [various dimensions]
h	hydraulic head, $[L]$
h_β	hydraulic head for phase β ($\beta = a, o, w$), $[L]$
h_c	capillary head (water-equivalent), $[L]$
h_{cap}	head at the well capture zone, $[L]$
h_d	Brooks-Corey displacement head, $[L]$
h_{ij}	capillary head for phase pair i,j ($ij = ao, aw, ow$), $[L]$
h_0	initial or baseline head, $[L]$
$h*$	scaled hydraulic head, $[L]$
i_w	slope of water table
$\mathbf{J}_\beta^i$	diffusive mass flux for chemical component i in phase β ($\beta = a, o, w$), $[ML^{-2}T^{-1}]$
$\mathbf{J}_{w,l}^i$	longitudinal (in the direction of flow) dispersive mass flux, $[ML^{-2}T^{-1}]$
$\mathbf{J}_{w,t}^i$	transverse (to the direction of flow) dispersive mass flux, $[ML^{-2}T^{-1}]$
K	hydraulic conductivity, $[LT^{-1}]$
K_β	hydraulic conductivity of phase β ($\beta = a, o, w$), $[LT^{-1}]$
K_d^i	soil-water distribution coefficient for chemical I, $[L^3M^{-1}]$
K_p^i	NAPL-water partition coefficient
K_s	saturated hydraulic conductivity, $[LT^{-1}]$
k	intrinsic permeability, $[L^2]$
k_{mt}^i	mass transfer coefficient $[LT^{-1}]$
$k_{mt}^i a$	mass transfer coefficient-interfacial area product, $[T^{-1}]$
k_{mt}^{i*}	lumped mass transfer rate coefficient, $[T^{-1}]$

$k_{r\beta}$	relative permeability for phase β ($\beta = a, o, w$)
$\overline{k}_{r\beta}$	average relative permeability for phase β ($\beta = a, o, w$)
k_{rnw}	relative permeability for non-wetting phase
k_{rw}	relative permeability for wetting phase
k_β	effective permeability for phase β ($\beta = a, o, w$)
l_f	location of DNAPL front, $[L]$
M	viscosity ratio
M_{wt}	average molecular weight of a fluid phase, $[M\ mol^{-1}]$
M_{wt}^i	molecular weight of chemical component i, $[M\ mol^{-1}]$
m	van Genuchten parameter
m_β	mass of phase β, $[M]$
m_β^i	mass of the chemical i in phase β, $[M]$
N_B	Bond number
N_{ca}	capillary number
N_T	total trapping number
n	van Genuchten parameter
N	total number of gas phase components
p	fluid pressure, $[ML^{-1}T^{-2}]$
p_β	pressure in phase β ($\beta = a, o, w$), $[ML^{-1}T^{-2}]$
$p_{\beta 0}$	pressure in phase β at the datum ($\beta = a, o, w$), $[ML^{-1}T^{-2}]$
p_a^i	gas partial pressure of chemical i, $[ML^{-1}T^{-2}]$
p_{vap}^i	vapor pressure, $[ML^{-1}T^{-2}]$
p_c	capillary pressure, $[ML^{-1}T^{-2}]$
p_{cd}	Brooks-Corey parameter for displacement pressure, $[ML^{-1}T^{-2}]$
p_{c0}	capillary pressure at the datum, $[ML^{-1}T^{-2}]$
p_{cij}	capillary pressure between fluid phase pair ij ($ij = ao, aw, ow$), $[ML^{-1}T^{-2}]$
p_d	Brooks-Corey displacement pressure, $[ML^{-1}T^{-2}]$
p_E	entry pressure for fracture, $[ML^{-1}T^{-2}]$
p_0	pressure at the datum, $[ML^{-1}T^{-2}]$

p_o characteristic pressure, $[ML^{-1}T^{-2}]$

p_{nw} non-wetting phase pressure, $[ML^{-1}T^{-2}]$

p_w wetting phase pressure, $[ML^{-1}T^{-2}]$

Q_o LNAPL recovery rate, $[L^3T^{-1}]$

Q_w water production rate, $[L^3T^{-1}]$

q_β specific discharge (Darcy velocity) of phase β ($\beta = a, o, w$), $[LT^{-1}]$

$\bar{q}$ mean specific discharge (Darcy velocity), $[LT^{-1}]$

R universal gas constant, $[ML^2T^{-2}mo\ |^{-1}k^{-1}]$

R_β^i retardation coefficient for phase β ($\beta = a, o, w$)

r_c radius of capture zone for a pumping well, $[L]$

r_{mt}^i rate of chemical mass transfer into the gas phase per unit volume of porous media

S_β saturation for phase β ($\beta = a, o, w$)

S_e normalized effective saturation

S_{et} effective total liquid saturation

S_{ew} normalized effective water saturation

S_{nwr} trapped saturation or irreducible non-wetting phase saturation

S_{or} residual NAPL saturation

S_{orv} residual NAPL saturation for unsaturated region

S_{ors} residual NAPL saturation for saturated region

S_t total liquid saturation

S_w wetting phase saturation

S_{wir} irreducible wetting phase or water saturation

$S*$ scaled saturation function

Sh' modified Sherwood number

T absolute temperature, $[K]$

T_{boil}^i temperature at boiling point of component I, $[K]$

u average velocity, $[LT^{-1}]$

V_o NAPL specific volume (NAPL volume in the soil per unit area in the horizontal plane), $[L^3]$

$V_{o,d}$ LNAPL specific volume at the destination point, $[LT^{-1}]$

$V_{o,s}$ LNAPL specific volume at the source point, $[L^3]$

v_β pore or interstitial velocity for phase β ($\beta = a, o, w$), $[LT^{-1}]$

v pore or interstitial velocity, $[LT^{-1}]$

w_s weighting function

X_T^i total mass fraction (mass of chemical per unit mass of dry (clean) soil or rock

X_β^i mass fraction: mass of chemical i per unit mass of the phase β

Z gas compressibility factor

z_{ij} air-oil interface elevation for fluid phase pair i,j ($ij = ao, aw, ow$), $[L]$

z_{fao} upper boundary of the LNAPL-air capillary fringe, $[L]$

z_u elevation of soil surface, $[L]$

Greek Symbols

α van Genuchten parameter, $[L^{-1}]$

α_l longitudinal dispersivity, $[L]$

α_t transverse dispersivity, $[L]$

β_{ow} scaling factor for pressure-saturation relationships

β_{ao} scaling factor pressure-saturation relationships

Γ relative mobility of a NAPL to water

λ Brooks-Corey pore size distribution parameter

λ_{crit} critical wavelength, $[L]$

λ_p perturbation wavelength, $[L]$

μ viscosity, $[ML^{-1}T^{-1}]$

μ_0 LNAPL to water viscosity ratio

ω angle of inclination

ϕ porosity

ρ density, $[ML^{-3}]$

ρ_b dry bulk density of soil or rock, $[ML^{-3}]$

ρ_g	density of the gas phase, $[ML^{-3}]$
ρ_{ro}	oil specific gravity (ratio of oil to water density)
ρ_β	phase density, $[ML^{-3}]$
$\Delta\rho$	density difference between water and NAPL, $[ML^{-3}]$
σ	interfacial tension, $[ML^{-3}]$
σ_{ij}	interfacial tension for phase pair (i,j), $[ML^{-3}]$
σ^*	macroscopic effective surface tension, $[MT^{-2}]$
τ	tortuosity factor in gas phase
τ_β	tortuosity factor in phase β
θ	contact angle
θ_w	wetting phase content
θ_{wir}	irreducible wetting phase content
χ_β^i	mole fraction: moles of chemical i per total moles of phase β

Indices and Qualifiers

a	air phase
f	DNAPL front
g	gas phase
i	chemical component
l	lower limit
nw	non-wetting phase
o	NAPL phase
s	solid phase
T	total
u	upper limit
w	wetting phase or water phase
β	gas, aqueous, NAPL, or solid phase
0	datum or initial

List of Figures

List of Tables

References

Abriola, L.M. (1984), *Multiphase Migration of Organic Compounds in a Porous Medium, Lecture Notes in Engineering, 8,* Springer-Verlag, Berlin.

American Petroleum Institute (1996), *In Situ Air Sparging,* American Petroleum Institute, Publ. 1628D, Washington, DC.

American Petroleum Institute (1996), *A Guide to the Assessment and Remediation of Underground Petroleum Releases,* American Petroleum Institute, Publ. 1628, Washington, DC.

Aziz, K. and A. Settari (1986), *Petroleum Reservoir Simulation,* Elsevier, London.

Baehr, A.L., G.E. Hoag, and M.C. Marley (1989), Removing volatile contaminants from the unsaturated zone by inducing advective air-phase transport, *J. Contam. Hydrol. 4,* 1-26.

Baehr, A.L., and G.E. Hoag (1988), A modeling and experimental investigation of induced venting of gasoline-contaminated soils, in *Soil Contaminated by Petroleum,* John Wiley and Sons, New York, 113-123.

Banerjee, S. (1984), Solubility of organic mixtures in water, *Env. Sci. Tech., 18,* 587-591.

Bear, J. (1972), *Dynamics of Fluids in Porous Media,* Dover Publications, New York.

Bear, J. (1979), *Hydraulics of Groundwater,* McGraw-Hill Publishing Company, New York.

Beckett, G.D. and S. Joy (2003), *Light Non-Aqueous Phase Liquid (LNAPL) Parameters Database, Version 2.0, Users Guide,* American Petroleum Institute, Publ. 4731, Washington, DC.

Bedient, P.B, H.S. Rifai and C.J. Newell (1999), *Ground Water Contamination: Transport and Remediation,* Prentice Hall, Englewood Cliffs, New Jersey.

Bejan, A. (1984), *Convection Heat Transfer,* John Wiley and Sons, New York.

Berg, L.L., K. Mansoor, C.M. Noyes, Z. Meir, R.W. Bainer, M.P. Maley, and S.D. Gregory, (2002), *Second Five-Year Review for the Lawrence Livermore National Laboratory Livermore Site,* UCRL-AR-147940, Lawrence Livermore National Laboratory, Livermore, California.

Bogacki, W. and Daniels, H. (1989), Optimal design of well location and automatic optimization of pumping rates for aquifer cleanup, in *Contaminant Transport in Groundwater,* edited by H. E. Kobus and W. Kinzelbach, A.A. Balkema Publishers, Rotterdam, 363-370.

Brooks, R.H. and A.T. Corey (1964), *Hydraulic Properties of Porous Media,* Hydrology Paper No. 3, Colorado State University, Fort Collins, Colorado.

Burdine, N.T. (1953), Relative permeability calculations from pore size distribution data, *Petroleum Transactions, AIME, 198,* 71-78.

Butts, M. B. (1996), *Flow of Miscible and Immiscible Hydrocarbons in Heterogeneous Porous Media,* Department of Hydrodynamics and Water Resources, Technical University of Denmark, Series Paper No. 61, Lyngby, Denmark.

Carsel, R.F. and R.S. Parrish (1988), Developing joint probability distributions of soil water retention characteristics, *Water Resour. Res., 24,* 755-769.

Carslaw, J.C. and J.C. Jaeger (1988), *Conduction of Heat in Solids*, Oxford University Press, Oxford, United Kingdom.

Charbeneau, R.J. (2000), *Groundwater Hydraulics and Pollutant Transport*, Prentice Hall, Upper Saddle River, New Jersey.

Charbeneau, R.J., R.T. Johns, L.W. Lake, and M.J. McAdams III (1999), *Free-Product Recovery of Petroleum Hydrocarbon Liquids*, American Petroleum Institute, Publ. 4682, Washington, DC.

Charbeneau, R.J., R.T. Johns, L.W. Lake, and M.J. McAdams III (2000), Free-product recovery of petroleum hydrocarbon liquids, *Ground Water Monit. Remediation, 20*, 147-158.

Chuouke, R.L., P. van Meurs, and C. van der Poel, (1959), The instability of slow, immicible, viscous liquid-liquid displacement in permeable media, *Pet. Trans., AIME, 216*, 188-194.

Cline, P.V., J.J. Delfino, and P.S.C. Rao (1991), Partitioning of aromatic constituents into water from gasoline and other complex solvent mixtures, *Environ. Sci. Tech., 25*, 914-920.

Compos, R., Jr. (1998), Multiple experimental realizations of dense nonaqueous phase liquid spreading in water saturated heterogeneous porous media, MS Thesis, Dept. Civil and Environmental Engineering, University of Colorado, Boulder.

Cooper, G.S., R.C. Peralta, and J.J. Kaluarachchi (1995), Stepwise pumping approach to improve free phase light hydrocarbon recovery from unconfined aquifers, *J. Contam. Hydrol., 18*, 141-159.

Corey, A.T. (1994), *Mechanics of Immiscible Fluids in Porous Media*, Water Resources Publications, Highlands Ranch, Colorado.

Davis, E.L. (1997), How heat can enhance in-situ soil and aquifer remediation: Important chemical properties and guidance on choosing the appropriate techniques, *EPA Ground Water Issue*, EPA/540/S-97/502, U.S. Environmental Protection Agency, Washington, DC.

Dawson, H.E., and P.V. Roberts (1996), Influence of viscous, gravitational, and capillary forces on DNAPL saturation, *Ground Water, 35*, 216-269.

Delshad, M., G.A. Pope and K. Sepehrnoori (1996), A compositional simulator for modeling surfactant enhanced aquiver remediation, 1: Formulation, *J. Contam. Hydrol. 24*, 247-267.

Environmental Protection Agency (1991), *Survey of Materials-Handling Technologies Used at Hazardous Waste Sites*, EPA/540/2-91/010, U.S. Environmetal Protection Agency, Washington, DC.

Environmental Protection Agency (1997), *Best Management Practices (BMPs) for Soils Treatment Technologies*, EPA/530/R-97/007, U.S. Environmental Protection Agency, Washington, DC.

Environmental Protection Agency (1998), *Remediation Case Studies: Groundwater Pump and Treat (Chlorinated Solvents)*, EPA/542/R-98/013, U.S. Environmental Protection Agency, Washington, DC.

Environmental Protection Agency (1999a), *Solidification/Stabilization Resource Guide*, EPA/542/B-99/002, U.S. Environmental Protection Agency, Washington, DC.

Environmental Protection Agency (1999b), *Hydraulic Optimization Demonstration for Groundwater Pump-and-Treat Systems, Volume I: Pre-Optimization Screening (Method and Demonstration*), EPA/542/R-99/011A, U.S. Environmental Protection Agency, Washington, DC.

Environmental Protection Agency (1999c), *Hydraulic Optimization Demonstration for Groundwater Pump-and-Treat Systems, Volume II: Application of Hydraulic Optimization*, EPA/542/R-99/011A, U.S. Environmental Protection Agency, Washington, DC.

Environmental Protection Agency (2003), *The DNAPL Remediation Challenge: Is There a Case for Source Depletion?*, EPA/600/R-03/143, U.S. Environmental Protection Agency, Washington, DC.

Fairbanks, T. M. (1993), Light non-aqueous phase fluid barriers in initially water saturated heterogeneous porous media, MS Thesis, University of Colorado, Boulder, Colorado.

Falta, R. W., C. M. Lee, S. E. Brame, E. Roeder, J. T. Coates, C. Wright, A.L. Wood, and C. G. Enfield (1999), Field test of high molecular weight alcohol flushing for subsurface NAPL remediation, *Water Resour. Res.*, *35*, 2095-2108.

Falta, R.W., K. Pruess, and D.A. Chesnut (1993), Modeling advective contaminant transport during soil vapor extraction, *Ground Water*, *31*, 1011-1019.

Farr, A.M., J. Houghtalen, and D.B. McWhorter (1990), Volume estimate of light non-aqueous phase liquids in porous media, *Ground Water,* *28*, 48-56.

Feenstra, S., Mackay, D.M., and J.A. Cherry (1991), A method for assessing residual NAPL based on organic chemical concentrations in soil samples, *Ground Water Monit. Rev.*, *11*, 128-136.

Fetter, C.W. (1999), *Contaminant Hydrogeology*, Prentice Hall, Englewood Cliffs, New Jersey.

Fischer, U., R. Schulin, M. Keller, and F. Stauffer (1996), Experimental and numerical investigation of soil vapor extraction, *Water Resour. Res.*, *28*, 3413-3427.

Freeman, H.M. and E.F. Harris (1995), *Hazardous Waste Remediation– Innovative Treatment Technologies*, Technomic Publ. Co., Landcaster, UK.

Gates, D.D. and R. L. Siegrist (1995), In situ chemical oxidation of trichloroethylene using hydrogen peroxide, *J. Environ. Engin.*, *121*.639-644.

Gavaskar, A.R., N. Gupta, B.M. Sass, R.J. Janosy, and D. O'Sullivan (1998), *Permeable Barriers for Groundwater Remediation*, Battelle Press, Columbus, Ohio.

Gelhar, L.W. (1993), *Stochastic Subsurface Hydrology*, Prentice Hall, Englewood Cliffs, New Jersey.

Gierke, J.S., N.J. Hutzler, and D.B. McKenzie (1992), Vapor transport in unsaturated soil columns: Implications for vapor extraction, *Water Resour. Res.*, *28*, 323-335.

Green, W.A. and G.A. Ampt (1911), Studies of soil physics, 1. The flow of air and water through soils, *J. Agr. Sci.*, *4*, 1-24.

Hall, R.A., S.B. Blake, and S.C. Champlin, Jr. (1984), Determination of hydrocarbon thickness in sediments using borehole data, in *Proceedings of the 4th National Symposium on Aquifer Restoration and Ground Water Monitoring*, National Water Well Assoc., Worthington, Ohio, 300-304.

Held, R. J. (1993), Investigation of fingering of dense nonaqueous phase liquids in saturated porous media, MS Thesis, University of Colorado, Boulder, Colorado.

Held, R.J. and T.H. Illangasekare (1995a), Fingering of dense nonaqueous phase liquids in porous media 1. Experimental investigation, *Water Resour. Res.*, *31*, 1213-1222.

Held, R.J. and T.H. Illangasekare (1995b), Fingering of dense nonaqueous phase liquids in porous media 2. Analysis and classification. *Water Resour. Res.*, *31*, 1223-1231.

Helmig, R. (1997), *Multiphase Flow and Transport Processes in the Subsurface*, Springer-Verlag, Berlin.

Hill, S. (1952), Channeling in packed columns, *Chem. Eng. Sci.*, *1*, 247-253.

Hiller, D.H. (1991), Performance characteristics of vapor extraction systems operated in Europe, in *Symposium on Soil Venting,* Houston, Texas, April 29-May 1, 1991, EPA/600/R-92/174, U.S. Environmental Protection Agency, Washington, DC, 193-202.

Hinchee, R.E., R.N. Miller, and P.C. Johnson (1995), *In Situ Aeration: Air Sparging, Bioventing, & Related Remediation Processes*, Battelle Press, Columbus, Ohio.

Ho, C.K. and K.S. Udell, (1992) An experimental investigation of air venting of volatile liquid hydrocarbon mixtures from homogeneous and heterogeneous porous media, *J. Contam. Hydrol.*, *11*, 291-316.

Ho, C.K., and K.S. Udell (1991), A mass transfer model for the removal of a volatile organic compound from heterogeneous porous media during vacuum extraction, *ASME Heat Transfer Geophysical Media, 172*, 55-62.

Ho, C.K., and S.W. Webb (1998), Review of porous media enhanced vapor-phase diffusion mechanisms, models, and data– Does enhanced vapor-phase diffusion exist?, *J. Porous Media, 1*, 71-92.

Hofstee, C., J.H. Dane, and W.E. Hill (1997), Threee-fluid retention in porous media involving water, PCE and air, *J. Contam. Hydrol.*, *25*, 235-247.

Homsey, G.M. (1987), Viscous fingering in porous media, *Ann. Rev. Fluid Mech.*, *19*, 271-311.

Hunt, J.R., N. Sitar, and K.S. Udell (1988), Nonaqueous phase liquid transport and cleanup 1. Analysis of mechanisms, *Water Resour. Res.*, *24*, 1247-1258.

Huntley, D., and G.D. Beckett (2002a), *Evaluating Hydrocarbon Removal from Source Zones and its Effect on Dissolved Plume Longevity and Concentration*, American Petroleum Institute, Publ. 4715, Washington, DC.

Huntley, D., and G.D. Beckett (2002b), Persistence of LNAPL sources: Relationship between risk reduction and LNAPL recovery, *J. Contam. Hydrol.*, *59*, 3-26.

Huntley, D., and G.D. Beckett (2004), *Evaluating Hydrocarbon Removal from Source Zones: Tools to Assess Concentration Reduction*, American Petroleum Institute, Washington, DC, in press.

Illangasekare, T.H. (1998), Flow and Entrapment of nonaqueous phase liquids in heterogeneous soil formations, in *Physical Nonequilibrium in Soils– Modeling and Application*, edited by H.M. Selim and L. Ma, Ann Arbor Press, Chelsea, Michigan., 417-435.

Illangasekare, T. H., D. N. Yates, and E. J. Armbruster (1995a), Effect of heterogeneity on transport and entrapment of nonaqueous phase waste products in aquifers: An experimental study, *J. Envir. Engin.*, *121*, 572-579.

Illangasekare, T.H., J.L. Ramsey, K.H. Jensen, and M. Butts (1995b), Experimental study of movement and distribution of dense organic contaminants in heterogeneous aquifers, *J. Contam. Hydrol.*, *20*, 1-25.

Imhoff, P.T and C.T. Miller (2000), Nonaqueous phase liquids in the subsurface: effects of nonaqueous-aqueous mass transfer limitations, in *Groundwater Pollution Control*, edited by K. Katsifarakis, Computational Mechanics Publications, Southampton,UK, 49-94, 2000.

Imhoff, P.T., P.R. Jaffé, and G.F. Pinder (1994), An experimental study of complete dissolution of a nonaqueous phase liquid in saturated porous media, *Water Resour. Res.*, *30*, 307-320.

Jackson, R.E. (2003), The evolution of DNAPL remediation practice, in *Chlorinated Solvent and DNAPL Remediation- Innovation Strategies for Subsurface Cleanup*, ACS Symposium Series No. 837, American Chemical Society, Washington, DC.

Jin, M., M. Delshad, V. Dwarakanath, D.C. McKinney, G.A. Pope, K. Sepehrnoori, C. Tilburg, and R.E. Jackson (1995), Partitioning tracer test for detection, estimation and remediation performance assessment of subsurface nonaqueous phase liquids, *Water Resour. Res.*, *31*, 1201-1211.

Johns, R.T., L.W. Lake, A.B. Obigbesan, L. Bermudez, M.R. Hassan, and R.J. Charbeneau (2003), Analytical solutions for free-hydrocarbon recovery using skimmer and dual-pump wells, *Ground Water Monit. Remediation, 23*, 97-106.

Johnson, P.C., C.C. Stanley, M.W. Kemblowski, D.L. Byers, and J.D. Colthart (1990a), A practical approach to the design, operation and monitoring of in situ soil-venting systems, *Ground Water Monit. Rev., Spring*, 159-178.

Johnson, P.C., M.W. Kemplowski, and J.D. Colthart (1990b), Quantitative analysis for the cleanup of hydrocarbon-contaminated soils by in-situ soil venting, *Ground Water*, *28*, 413-429.

Johnson, R.L. and J.F. Pankow (1992), Dissolution of dense chlorinated solvents into groundwater. 2. Source functions for pools of solvent, *Environ. Sci. Technol., 26*, 896-901.

Jury, W.A., W.F. Spencer, and W.J. Farmer (1983), Behavior assessment model for trace organics in soil, I. Model description, *J. Environ. Qual., 12*, 558-564.

Jury, W.A., W.F. Spencer, and W.J. Farmer (1984), Behavior assessment model for trace organics in soil, III. Application of screening model, *J. Environ. Qual., 13*, 567-572.

Kaluarachchi, J.J. and R. T. Elliott. (1995), Design factors for improving the efficiency of free-product recovery systems in unconfined aquifers, *Ground Water*, *33*, 909-916.

Kemblowski, M.W. and C.Y. Chiang (1990), Hydrocarbon thickness fluctuations in monitoring wells, *Ground Water*, *28*, 244-252.

Knauss K.G., R.D. Aines, M.J. Dibley, R.N. Leif and D.A. Mew (1997), *Hydrous Pyrolysis/Oxidation: In-Ground Thermal Destruction of Organic Contaminants*, LNL UCRL-JC-126636, Lawrence Livermore National Laboratory, Livermore, California.

Kram, M.L., A.A. Keller, J. Rossabi, and L.G. Everett (2001), DNAPL characterization methods and approaches, Part 1: Performance comparisons, *Ground Water Monit. Remediation, Fall*, 109-123.

Kram, M.L., A.A. Keller, J. Rossabi, and L.G. Everett (2002) DNAPL characterization methods and approaches, Part 2: Cost comparisons, *Ground Water Monit. Remediation, Winter*, 46-61.

Kueper, B. H. and D. B. McWhorter (1991), The behavior of dense, nonaqueous phase liquids in fractured clay and rock, *Ground Water*, *29*, 716-728.

Kueper, B.H. and E.O. Frind (1991), Two-phase flow in heterogeneous porous media 1. Model development, *Water Resour. Res., 27*, 1049-1057.

Lee, S., M. Hagwall, J. Delfino, and P.S.C. Rao (1992), Partitioning of polycyclic aromatic hydrocarbons from diesel fuel into water, *Env. Sci. Tech.*, *26*, 2104-2109.

Leij, F., W.J. Alves, M.Th. van Genuchten, and J.R. Williams (1996), *The UNSODA Unsaturated Soil Hydraulic Database, User's Manual Version 1.0*, EPA/600/R-96/095, U.S. Environmental Protection Agency, Washington, DC.

Lenhard, R.J. and J.C. Parker (1987), Measurement and prediction of saturation-pressure relationships in three phase porous media systems, *J. Contam. Hydrol.*, *1*, 407-424.

Lenhard, R.J. and J.C. Parker (1990), Estimation of free product hydrocarbon volume from fluid levels in monitoring wells, *Ground Water*, *28*, 57-67.

Lenormand, R. (1985), Difference mecanismes de displacments visqueux et capillaires en milieu poreux: Diagram de phase, C.R. Acad. Sci. Paris Ser. II, *310*, 247-250.

Leverett, M.C. (1941), Capillary behavior in porous solids, *Trans. AIME Pet. Eng.*, *142*, 152-169.

Londergan, J. T., H.W. Meinardus, P.E. Mariner, R.E. Jackson, C.L. Brown, V. Dwarakanath, G.A. Pope, J.S. Ginn, S. Taffinder (2001), DNAPL removal from a heterogeneous alluvial aquifer by surfactant-enhanced aquifer remediation, *Ground Water Monit. Remediation*, Fall, *21*, 57-67.

Looney, B.B., and R.W. Falta (2000), *Vadose Zone Science and Technology Solutions*, Battelle Press, Columbus, Ohio.

Mackay, D. M., W. Y. Shiu, A. Maijanen, and S. Feenstra (1991), Dissolution of non-aqueous phase liquids in groundwater, *J. Contam. Hydrol.*, *8*, 23-42.

Marinelli, F. and D.S. Durnford (1996), LNAPL thickness in monitoring wells considering hysteresis and entrapment, *Ground Water*, *34*, 405-414.

Mariner, P.E., M. Jin, and R.E. Jackson (1997), An algorithm for the estimation of NAPL saturation and composition from typical soil chemical analyses, *Ground Water Monit. Remediation*, 17, 122-129.

Mayer, A.S. and C.T. Miller (1996), The influence of mass transfer characteristics and porous media heterogeneity on nonaqueous phase liquid dissolution, *Water Resour. Res.*, *32*, 1551-1567.

Mayer, A.S., Zhong, L. and G. Pope (1999), Measurement of mass transfer rates for surfactant-enhanced solubilization of nonaqueous phase liquids, *Environ. Sci. Technol.*, *33*, 2965-2972.

McWhorter, D.B. and B. H. Kueper (1996), Mechanics and mathematics of the movement of dense non-aqueous phase liquids (DNAPLS) in porous media, in *Dense Chlorinated Solvents and other DNAPLs in Groundwater*, edited by J.F. Pankow and J.A. Cherry, Waterloo Press, Waterloo, Ontario, Canada, 89-128.

Mercer, J.W. and Cohen, R.M. (1990), A review of immiscible fluids in the subsurface: properties, models, characterization and remediation, *J. Contam. Hydrol.*, *6*, 107-163.

Miller, C.T., Poirier-McNeill, M.M. and A.S. Mayer (1990), Dissolution of trapped non-aqueous phase liquids: Mass transfer characteristics, *Water Resour. Res.*, *26*, 2783-2796.

Millington, R. J. and J. P. Quirk (1961), Permeability of porous soils, *Trans. Faraday Soc.*, *57*, 1200-1207.

Morrow, N.R. and B. Songkran (1981), Effect of viscous and buoyancy forces on non-wetting phase trapping in porous media, in *Surface Phenomena in Enhanced Oil Recovery*, edited by D.O. Shah, Plenum Press, New York, 387-411.

Mott, H.V. (1995), A model for determination of the phase distribution of petroleum hydrocarbons at release sites, *Ground Water Monit. Remediation*, *15*, 157-167.

Mualem, Y. (1976), A new model for predicting the hydraulic conductivity of unsaturated porous media, *Water Resour. Res.*, *12*, 513-522.

Naftz, D., Morrison, S.J., Fuller, C.C., and Davis, J.A., editors, (2002), *Handbook of Groundwater Remediation Using Permeable Reactive Barriers – Applications to Radionuclides, Trace Metals, and Nutrients*, Academic Press, San Diego, California.

National Research Council (1994), *Alternatives for Ground Water Cleanup*, National Academy Press, Washington, DC.

National Research Council (1997), *Barrier Technologies for Environmental Management*, National Academy Press, Washington, DC.

Nelson, M. D., B. L. Parker, T. A. Al, J. A. Cherry, and D. Loomer (2001) Geochemical reactions resulting from in situ oxidation of PCE-DNAPL by $KMnO_4$ in a sandy aquifer, *Envir. Sci. Technol.*, *35*, 1266-1275.

Nyer, E.K., D.F. Kidd, P.L. Palmer, T.L. Crossman, S. Fam, F.J. Johns II, G. Boettcher, and S.S. Suthersan (1996), *In Situ Treatment Technology*, CRC Lewis Publishers, Boca Raton, Florida.

Ogata, A. (1970), *Theory of Dispersion in a Granular Medium*, U.S. Geological Survey Professional Paper 411-I, U.S. Geological Survey, Reston, Virginia.

Pankow, J.F. and J.A. Cherry (1996), *Dense Chlorinated Solvents & Other DNAPLs in Groundwater: History, Behavior, and Remediation*, Waterloo Educational Services Inc., Rockwood, Ontario, Canada.

Parker, B. L., R. W. Gillham and J. A. Cherry (1994), Diffusive disappearance of immiscible-phase organic liquids in fractured geologic media, *Ground Water*, *32*, 805-820.

Parker, J.C., D.W. Waddill and J.A. Johnson (1995), *UST Corrective Action Technologies: Engineering Design of Free Product Recovery Systems*, EPA/600/SR-96/031, U.S. Environmental Protection Agency, Washington, DC.

Parker, J.C., R.J. Lenhard and Kuppusamy, T. (1987), A parametric model for constitutive properties governing multi-phase flow in porous media, *Water Resour. Res.*, *23*, 618-624.

Parker, J.C. (1989), Multiphase flow and transport in porous media, *Rev. Geophysics*, 27, 311-328.

Pastrovich, T.L., Y. Baradat, R. Barthel, A. Chiarelli, and D.R. Fussell (1979), Protection of Groundwater from Oil Pollution, CONCAWE, Report 3/79. The Hague, Netherlands.

Pennell, K.D., G.A. Pope, and L.M. Abriola (1996), Influence of viscous and buoyancy forces on the mobilization and of residual tetrachloroethylene, *Environ. Sci. Technol.*, *30*, 1328-1335.

Pennell, K.D., M. Jin, L.M. Abriola, and G.A. Pope, (1994), Surfactant enhanced remediation of soil columns contaminated by residual tetrachloroethylene. *J. Contam. Hydrol.* *16*, 35-53.

Poulsen, M.M. and B.H. Kueper (1992), A field experiment to study the behavior of tetrachloroethylene in unsaturated porous media, *Environ. Sci. Technol.*, *26*, 889-895.

Powers, S. E., L. M. Abriola, and W. J. Weber, Jr. (1994), An experimental investigation of NAPL dissolution in saturated subsurface systems: Transient mass transfer rates, *Water Resour. Res.*, *30*, 321-332.

Prausnitz, J.M., R.N. Lichtenthaler, and E. G. de Azevedo (1986), *Molecular Thermodynamics of Fluid-Phase Equilibria*, Prentice-Hall, Englewood Cliffs, New Jersey.

Rao, P. S. C., M. D. Annable, R. K. Sillan, D. Dai, K. Hatfield, W. D. Graham, A. L. Wood, and C. G. Enfield (1997), Field-scale evaluation of in situ cosolvent flushing for enhanced aquifer remediation, *Water Resour. Res.*, *33*, 2673-2686.

Reible, D.D., T.H. Illangasekare, D. Doshi and M.E. Malhiet (1990), Infiltration of immiscible contaminants in the unsaturated zone, *Ground Water*, *28*, 685-692.

Rumer, R. and M.E. Ryan (1995), *Barrier Containment Technologies for the Environmental Remediation Applications*, John Wiley & Sons, New York.

Sabatini, D.A., R.C. Knox, J.H. Harwell (1996), *Surfactant-Enhanced DNAPL Remediation: Surfactant Selection, Hydraulic Efficiency, and Economic Factors*, EPA/600/S-96/002, U.S. Environmental Protection Agency. Washington, D.C.

Sale, T. (2001), *Methods for Determining Inputs to Environmental Petroleum Hydrocarbon Mobility and Recovery Models*, American Petroleum Institute, Publ. 4711, Washington, DC.

Schwille, F. (1988), Dense chlorinated solvents in porous and fractured media, translated by J. F. Pankow, Lewis Publishers, Chelsea, Michigan.

Seagren, E.A., B.E. Rittmann, and A.J. Valocchi (2002), Bioenhancement of NAPL-pool dissolution: Experimental evaluation, *J. Contam. Hydrol.*, *55*, 57-85.

Seol, Y., Schwartz, F. W. and S. Lee (2001), Oxidation of binary DNAPL mixtures using potassium permanganate with a phase transfer catalyst, *Ground Water Monit. Remediation*, *21(2)*, 124-133.

Shan, C., R.W. Falta, and I. Javandel (1992), Analytical solutions for steady state gas flow to a soil vapor extraction well, *Water Resour. Res.*, *28*, 1105-1120.

Simpkin T., T. Sale, B. Kueper, M. Pitts, and K. Wyatt (1999), *Surfactants and Cosolvents for NAPL Remediation, A Technology Manual*, Lewis Publishers, Boca Raton, Florida.

Sontag, R.E., and G. Van Wylen (1982), *Introduction to Thermodynamics*, John Wiley and Sons, New York.

Spooner, P., Wetzel, R., Spooner, C., Furman, C., Tokarshi, E., Hunt, G., Hodge, V., and Robinson, T. (1985), *Slurry Trench Construction for Pollution Migration Control*, Noyes Publications, Park Ridge, New Jersey.

Stephens, D. B., J. A. Kelsey, M. A. Prieksat, M. G. Piepho, C. Shan and M. D. Ankeny (1998), DNAPL migration through a fractured perching layer, *Ground Water*, *36*, 605-610.

Tiab, D. and E.C. Donaldson (1996), *Petrophysics*, Gulf Publishing Company, Houston, Texas.

van Dam, J. (1967), The migration of hydrocarbons in a water-bearing stratum, in *The Joint Problems of the Oil and Water Industries*, edited by P. Hepple, Institute of Petroleum, London, 55-96.

van Genuchten, M.Th. (1980), A closed-form equation for predicting the hydraulic conductivity of unsaturated soils, *Soil Sci. Soc. Am. J.*, *44*, 892-898.

Weaver, J.W., R.J. Charbeneau and B.K. Lien (1994), A screening model for non-aqueous phase liquid transport in the vadose zone using Green-Ampt and kinematic wave theory, *Water Resour. Res.*, *30*, 93-105.

White, M.D. and M. Oostrom, (2000), *Subsurface Transport Over Multiple Phases, STOMP V.2.0 User's Guide*, PNNL Report 12034, Pacific Northwest National Laboratory, Richland, Washington.

Wickramanayake, G.B., A.R. Gavaskar, and A.S.C. Chen (2000), *Chemical Oxidation and Reactive Barriers: Remediation of Chlorinated and Recalcitrant Compounds (C2-6)*, Batelle Press, Columbus, Ohio.

Wickramanayake, G.B. and R.E. Hinchee (1998), *Physical, Chemical, and Thermal Technologies: Remediation of Chlorinated and Recalcitrant Compounds (C1-5)*, Batelle Press, Columbus, Ohio.

Wilkins, M.D., L.M. Abriola, and K.D. Pennell (1995), An experimental investigation of rate-limited nonaqueous phase liquid volatilization in unsaturated porous-media: Steady state mass transfer, *Water Resour. Res., 31*, 2159-2172.

Wilson, D. (1995), *Modeling of In Situ Techniques for Treatment of Contaminated Soil: Soil Vapor Extraction, Sparging, & Bioventing*, Technomic Publishing Co., Lancaster, Pennsylvania.

Wilson, J.L., S.H. Conrad, W.R. Mason, and W. Peplinski (1990), *Laboratory Investigation of Residual Liquid Organics from Spills, Leaks, and the Disposal of Hazardous Wastes in Groundwater*, EPA/600/6-90/004, U.S. Environmental Protection Agency, Washington, DC.

Index

About the Authors

Dr. S. Majid Hassanizadeh is Professor of Hydrogeology in the Department of Earth Sciences of Utrecht University, The Netherlands. He received his Ph.D. from Princeton University in 1979.

Dr. Alex Mayer is a Professor in the Department of Geological & Mining Engineering & Sciences, Michigan Technological University. Dr. Mayer's Ph.D. in Environmental Engineering is from the University of North Carolina at Chapel Hill, 1992.

Printed and bound by CPI Group (UK) Ltd, Croydon, CR0 4YY

27/10/2024

14580242-0001